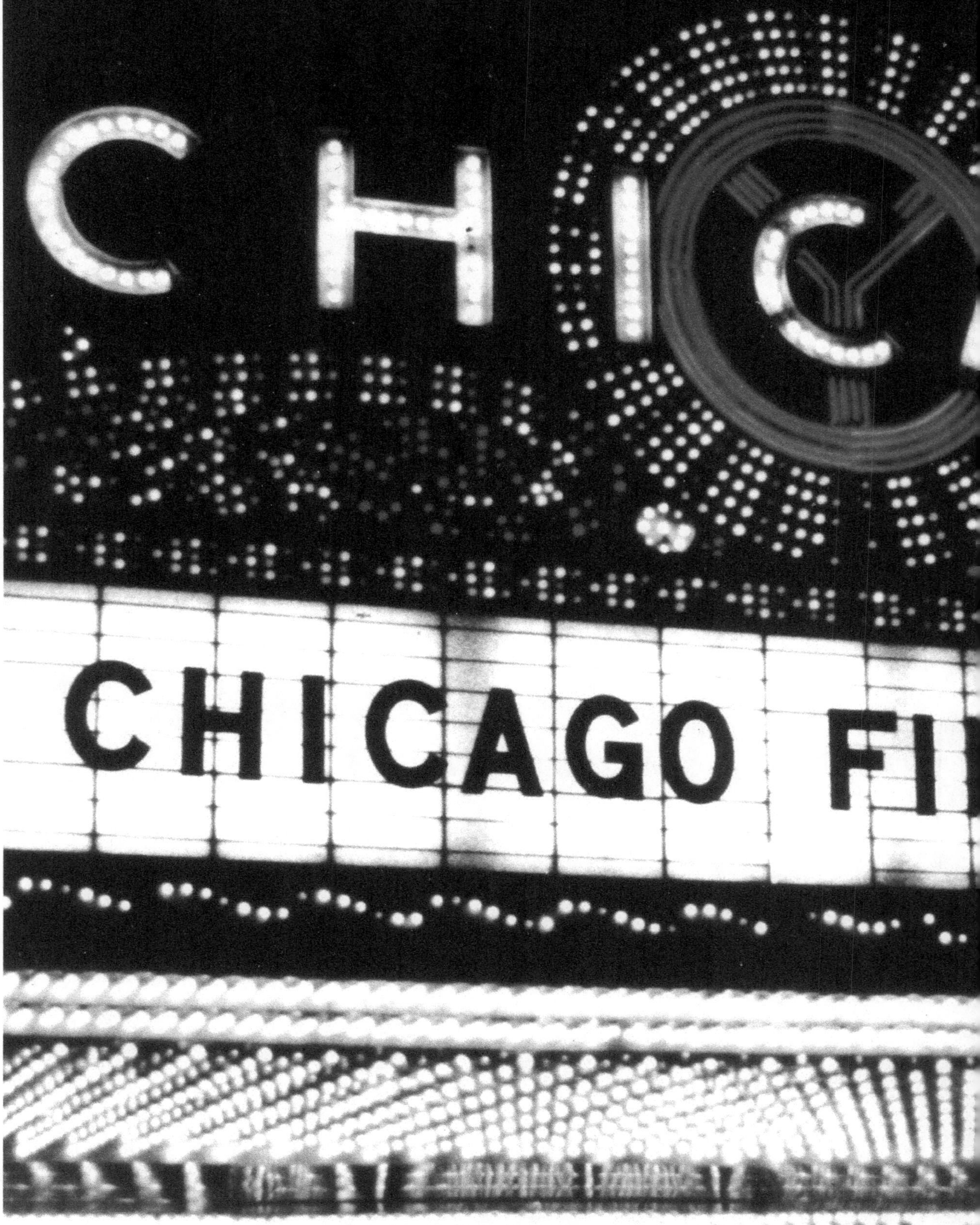
CHICAGO

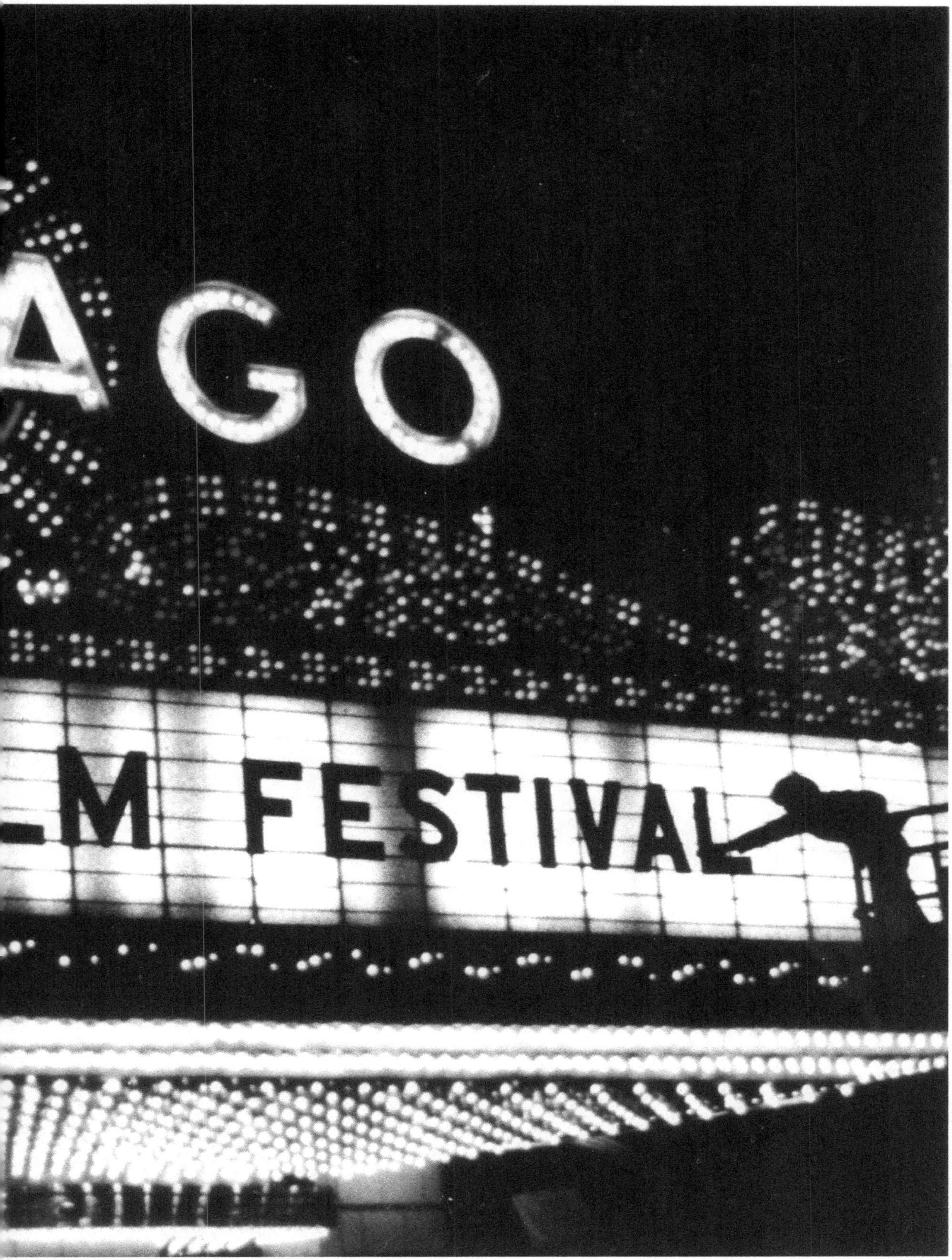
AGO
M FESTIVAL

Starstruck

Starstruck

How I Magically Transformed Chicago into Hollywood for More Than Fifty Years

By Michael Kutza

BearManor Media
2022

Starstruck

© 2022 *Michael Kutza*

All rights reserved.

Published in the United States of America by:

BearManor Media

4700 Millenia Blvd.
Suite 175 PMB 90497
Orlando, FL 32839

bearmanormedia.com

Printed in the United States.

Typesetting and layout by BearManor Media

Photo Credits: Cover
(Left} Anna Nicole Smith: Steve Arazmus
(Right) Spielberg/Cruise: Robert Dowey
(Lower left) Sophia Loren: Matt Gilson
(Lower right) Jack Lemmon: Steve Arazmus

Photo Credit: Back Cover
Photo of Michael Kutza: Tim Klein

ISBN—978-1-62933-957-3

Starstruck

- "Michael Kutza knew how to make a film festival informative, interesting, and a lot of fun. He is a great storyteller and hosted some of the biggest and best throughout his many years leading the Chicago festival. His book is a great read."

  -**Michael Douglas**, Oscar®-winning actor and producer of films including *Wall Street, One Flew Over the Cuckoo's Nest,* and *Romancing the Stone*

- "Michael has been strong and generous for years and years, pushing us all into his world of cinema to make us grow integrity, justice, and dignity – and to become better human beings."

  -**Geraldine Chaplin**, award-winning actress of films including *Doctor Zhivago* and *Nashville*

- "Michael has made me so happy and inspired through many meetings and film get-togethers. He is a teacher, a friend, and such a wonderful, sunny soul. Thank you for being, Michael."

  -**Liv Ullmann**, Oscar®-winning actress of films including *Persona* and *Cries & Whispers*

- "Michael dreamt big. His vision of bringing together international artists to celebrate cinema in Chicago was groundbreaking. Passion, tenacity and love for film got him there first!"

  -**Paula Wagner**, film executive and producer of *Marshall* and *Mission: Impossible I, II and III*

- "Anyone who has met Michael knows everything about the brilliant founder of the prestigious Chicago Festival, a friend of all the biggest movie stars who knows how to move around the world with humor, elegance, and self-irony."

  -**Roberto Cicutto**, film producer, former head of CINECITTA Studios and President of Venice Biennale

For Victor and Paula

TABLE OF CONTENTS - STARSTRUCK

FOREWORD

By David Robinson

This book is a surprise to me. Though I have been, I think, a close friend of the author-subject for a very long time, I have now discovered there was so much I did not know. Michael Kutza has known *everybody* and had a positive influence on many of those whose work was in film, thanks to the prestige of the international festival he created with the same boyish enthusiasm that had inspired his toy theaters, not so long before.

When Michael had the wild idea of creating a film festival in North America (Chicago) at a time when film festivals were still a fairly new idea, and even Cannes and Venice were still growing and learning, he sought out a partner in a legendary figure from the forgotten age of silent films. She was a woman whose intelligence and vitality made her an ongoing force in cinema society and politics.

Colleen Moore and her veteran beau, the master director of classic Hollywood, King Vidor, opened every door while at the same time insisting on adding five years to Michael's official age to bring him greater credibility.

I would love to have known him in the Colleen days, and also to have met that formidable lady. I met Michael much later, but when I asked him when and where that happened, he was peculiarly snappy ("What's *that* have to do with anything?" he fumed) before suggesting it was during one of the more indecorous moments in his life. It involved the charismatic and mischievous Albert Johnson, whom I never really knew. He even suggested it was the "What shall we do with the drunken sailor?" saga, which figures with colorful indecency in his story, but which occurred long before we met.

In any event, I believe it was in the early 1970s when a rather prissy Englishman who worked for the British Council said to me in Berlin, "You should meet a man called Michael Kutza from Chicago. He's funny." I did meet him, and he was indeed funny. We took to each other immediately. We both went to the same festivals and saw the same films. We sat together because we both like to be near the screen, and I always kept a place for him because he would only just get in as the lights went down. We appeared so inseparable (we ate together as well) that I know a lot of kindly colleagues suspected we were *A Couple*. Think what they might, there was never one moment of romantic attachment in our half century or so of friendship. We were friends, and that was that.

Latterly, to save money, we would share a flat at Cannes during that annual festival and keep house together. I went out near daybreak to get the bread, and he fried up unhealthy delights. We would throw parties for friends and clients and were left with stacks of uneaten pizzas. It was marvelous. We also made sure that doing our jobs and dutifully seeing all the films was fun, and rewardingly and permanently companionable.

Counting Michael Kutza as a friend has been one of the richest rewards of my long life. And I think you'll agree that his book makes for fascinating reading.

David Robinson is the former film critic for THE FINANCIAL TIMES, THE TIMES OF LONDON and other journalistic posts, He is also the former Director of the EDINBURGH INTERNATIONAL FILM FESTIVAL and for-mer Director and now Director Emeritus of LE GIONRNATE DEL CINEMA MUTO (Pordenone Silent Film Festival). He is also the author of the book, CHAPLIN: His Life and Art.

INTRODUCTION

"A CINEMATIC LIFE"
by John Russell Taylor

When it comes to film festival directors, I've known them all, and believe me, whatever divinities shaped Michael Kutza promptly broke the mold.

Festival directors come in all shapes and sizes, from ivory tower intellectuals to shamelessly exhibitionistic showmen, but Michael is unmistakably one of a kind. Though he was trained as a designer, and is brilliant at it, as the graphics of the Chicago International Film Festival have consistently attested, you cannot imagine that he would be anything else but dedicated heart and soul to film.

Who am I to be pontificating? Well, in the '60s when I first met Michael, I was film critic of the *London Times*. He invited me to be on the festival jury in 1971 and I've attended nearly all of the festivals since. In the '70s, I was a professor in the film division at USC in Los Angeles.

I befriended and wrote the first biography of Alfred Hitchcock, the only one written with his complete cooperation. In the '80s and '90s, I was back in London editing the monthly magazine *Films and Filming.* About half of the hundred or so books I've written are on film subjects. So rightly or wrongly, where Michael is concerned, I know what I'm talking about.

The thing about Michael is that, talking about him (and thousands do), you will inevitably find yourself speaking in superlatives, often mutually exclusive. Descriptions that spring to mind are shyest, boldest, funniest, most serious, most demanding, and most acerbic. Consequently, he was also the most maddening and most delightful of all film festival directors.

I'm proud to say that he is probably my best friend in the world, so don't expect me to be cool and measured in my portrait of him. His knowledge of cinema is encyclopedic, his enthusiasm limitless. See him operating at another festival (it was his decision early on that his festival, coming late in the year, should be, among other things. a festival of festivals, so it is interested in quality rather than exclusivity) and you realize how passionately he pursues the best in world cinema, the most exciting new talents.

If the obsession was not in his genes, it must have kicked in pretty early on. The younger son of a Polish father and an Italian mother, both successful doctors, he had a more international background than most of his school fellows and has always been the most effectively cosmopolitan American I know. He was immediately at home virtually anywhere in the world, though at the same time intensely loyal to his Chicago roots. Indeed, his initial motivation to launch a film festival was to bring the best of world cinema to his fellow Chicagoans.

As the festival heads toward its diamond jubilee, Michael can chalk up a unique success in that area. I have said that he is the shyest of festival directors. He is a brilliant public speaker and vocal advocate for his enthusiasms, but I think he does not believe in his own skills. While most festival directors love to bask in the limelight, he much prefers to stay hard at work behind his desk.

Genius in filmmakers both fascinates and intimidates him. (He was himself a prize-winning student filmmaker, but you have to drag the information out of him.) However, his vast knowledge and passionate appreciation have brought an amazing number of classic film stars and directors to the festival for in-depth onstage interviews, along with the cream of the crop of newcomers, many of whom were discovered by and launched at the festival. (Think Martin Scorsese, Oliver Stone, Taylor Hackford, Tom Cruise, John Cusack, Helen Mirren, Mike Leigh...the list goes on and on.)

Have I done justice to the glamour that hangs around Michael? Probably not. He is certainly the best looking of all festival directors I have known, and the most impressive public figure. He is also, I think, the only one in the world to have a street named after him.

When Michael retired from the festival, the Chicago scene was poorer by at least one of its brightest lights.

John Russell Taylor is the former film critic for THE TIMES OF LONDON and the author of 94 books on film and art, including HITCH: The Life & Times of Alfred Hitchcock, *the official biography of Alfred Hitchcock.*

1 - WEST SIDE STORY

Growing up on the West Side of Chicago, I had one driving ambition: to tell stories and show people something bigger than they had ever seen.

It all started when I nailed a white bedsheet to the wall in the basement to show movies and called in all the neighborhood kids to see them. I was like a producer with a captive audience. It was the same with my little puppet shows. With those, I may have only had an audience of one – the housekeeper – but she would see something fabulous. And one person was all I needed.

All of these moments were bigger than me.

I also knew that I wasn't going to become a doctor like my parents, though that dream would die hard – for them, not for me. However, I'm getting ahead of myself, something I have a nasty habit of doing.

As intimated, I grew up in a world of doctors. My mother, my father, my aunt, my uncle, even my cousins where all fucking doctors. Every damn one! You can see where this is leading. It's something I had zero interest in. Z-E-R-O. But if you're in a family of doctors, it's expected – no, *demanded* – that you become one, too. Consequently, I had to go through the motions of pretending I was destined for medical school and a career in medicine.

I wound up getting thrown out of three or four colleges because I didn't want to be a doctor and they could tell. It seems they really wanted you to study when you were in school. Who knew?

But let's travel back in time still further.

One morning when I was three years old, I was awakened by a big racket coming from downstairs. I slept on the second floor in my grandmother's room. They were moving out all of the furniture on the first floor to put it in storage. I sat and watched, confused.

They moved in a long black box and placed it in the living room. For seven days and nights, people would come to the house to view Grandma in a casket in the living room, because we Italians love to look at our dead relatives.

From that day forward, my young life would forever change, and not in a good way. I couldn't go near the living room. I couldn't go back to my bedroom because it was hers. I was terrified of even being in that house. I had already seen enough films at my young age to know that Grandma wasn't really dead and would be coming out of that box at night to haunt the house (and me in particular).

I retreated to the basement where I had my puppets and watched old movies my dad had collected. With the lady in the box gone forever, my folks hired a series of African American housekeepers to raise me. I quickly became the man of the house as a result.

I had to be independent from an early age because I was never really a child. Clearly, I never really had what you would consider a "normal" childhood. I had those puppets and built elaborate stages and sets to put on shows for the maid. My folks didn't care. They were dedicated physicians day and night at their respective offices and hospitals. I was left to fend for myself and fend I did. I became an independent mini adult early on.

Mind you, my shows got better and better. One Sunday, I actually convinced my parents to watch one – and it predictably wowed them. That was so cool that I still recall it vividly.

My mother, a rather large woman, specialized in obstetrics and gynecology. She was a big deal at her hospital Mother Cabrini Memorial in Chicago's Italian neighborhood. Dad, a good-looking ladies' man, was a surgeon and the president of his hospital Saint Mary's in the Polish district.

It seems that I had a brother, but he was ten years older and always gone, so I barely had any interaction with him. He was away at college, in the Army, then off and married. I never got to know him aside from the last few years of his life. I was basically an only child in my mind.

I decided one day that I must have been adopted. My parents seemed too old to have conceived me at their age. Then one day, I met a woman who had assisted at my birth. Mrs. House was a nurse at the hospital. Yes, House like the famed TV doc of the same name.

"The day that your mother came in for her delivery, I asked her, 'Where is the patient?'," Mrs. House recalled.

"I'm the patient," my mother is said to have replied.

My mother's size had hidden the fact that she was pregnant and about to give birth to me, as the story went. So, this woman whom I occasionally saw in the house really was my mother, it seemed. I further confirmed it by later locating birth records, footprints, date of birth, time of birth, and photo. So much for my airtight adoption theory.

Moviemaking came to me naturally, yet in an odd way. My dad collected old silent films and some sound features he could get his hands on as entertainment for me in the basement, which I like to call the original mancave. (Kidcave?) Mom actually shot movies. She had a 16-millimeter camera that she took with her on lady doctor trips all over the world. She would record these films of her trips and doctor conventions, then come back home, hand them to me and say, "Make these into something, the lady doctors will be coming over Sunday."

That was my cue to get to work.

I started making movies with music, special effects, fabulous stuff for an audience. I think I was all of eight years old by this time. I was old enough, however, to go to the neighborhood movie theater and watch double features every Saturday. Horror films, science fiction, musicals, and all of those great serials. I was all over them.

I not surprisingly grew a little starstruck back then. My mother tried to keep me busy beyond my flashy basement shows. She had one of her medical patients take me to a nightclub. It was a place like Las Vegas, but in Chicago: the Chez Paree.

I was maybe 12 when this happened. Suddenly, I was meeting Jimmy Durante, Sophie Tucker, Dean Martin, and Jerry Lewis in a place filled with real showgirls. It was all just impossibly exciting and glamorous. Mind you, I never asked my mother's patient about her connection with the club but assumed she was a girlfriend of the boss. All I knew was that this lady had an in and knew everybody.

One thing I quickly learned was that you're never supposed to ask how or why anything was happening. I just enjoyed the access. Through Chez Paree, the showbiz bug bit and grew inside me big time.

There was one other interesting aspect of the Chez: a lounge with a live radio broadcast emanating from there with the stars every night. I would listen to it at home before I went to sleep. Who would have thought that I'd often be on that show later in life and bring the visiting stars at my film festival to the show? Talk about full circle.

Besides showgirls at the Chez and my puppet shows in the basement at home, as I already mentioned, I spent a lot of time watching movie double features at my two neighborhood theaters. It was mostly horror and sci-fi films until musicals hit big and then 3-D came along. I got hooked. Man, did I ever. I think I watched *House of Wax* five times.

Everything really changed in 1953 when my cousin took me to a theater in downtown Chicago to see something new and revolutionary. We went to the Bismarck Theatre (which is now called the Cadillac Palace Theatre), a legitimate house that was turned into a movie palace for this very special event.

I was met with this giant red curtain wrapping around the whole theater. The main floor had about 1,200 seats, and the red curtain must have been two stories high and filled my peripheral vision in width. The program began as the curtains parted maybe a quarter of the way. A man appeared onscreen in black and white. His name was Lowell Thomas. I guess he'd been a famous radio commentator or something. He uttered the momentous line, "Ladies and gentlemen, THIS IS CINERAMA!"

The curtain continued to part and kept parting and parting and wrapping around me, and suddenly there was a rollercoaster ride depicted on the full screen along with this new thing called stereophonic sound. I was totally blown away and went back maybe five times. It was expensive, but I had to see it over and over again. After a while, I grew tired of the travelogue theme of the film, but the showmanship and the whole situation was simply captivating.

What proved ironic was the fact that showman Michael Todd and his son, Michael Todd Jr., directed this success story, yet within maybe ten years they would effectively destroy the three-projector system of Cinerama with their own new widescreen innovation, Todd-AO. Later, CinemaScope and all of the other systems replicating this complicated three-projector system came in. But it was Cinerama that changed how we saw the movies forever.

I was very impressed with the Todds and how they presented the movies to audiences. They were trendsetters. Later, Michael Jr. did something that was quite amusing, a film called *The Scent of Mystery* that was presented in SMELL-O-VISION. Why not involve *all* of the senses, right?

it only lasted maybe a year or two, but Todd specially built the theater just to do it here in Chicago and in New York. It was really fun and silly. He installed little tubing under all of the seats that would pump the various smells out according to what was being depicted on the screen.

In between the smells, a large, loud burst of fresh air had to be pumped in to clear out the previous scent. It was annoying, but necessary. A clever part of the film involved a mysterious woman would who appear throughout and was always accompanied by a certain perfume fragrance. You never saw her face, but you knew when she was around because they had trained us to know her smell.

At the end of the film, the actress turned to the audience and we saw that it was Elizabeth Taylor, Todd's wife at the time. It's funny when you consider that years later, Liz would be one of the first stars to come out with her own fragrance, White Diamonds.

One thing I had not been exposed to yet was foreign films. I was still mostly watching the regular double bills on the West Side and really going back to see the Flash Gordon and Dick Tracy serials every weekend. The foreign stuff didn't come into my life until 1958, with Ingmar Bergman's *Seventh Seal*. It was shot in Swedish with English subtitles and playing at one of our three little Chicago art houses.

Back in those days, a foreign film like this would play in that same theater for a year. A year! The other two art houses had films for runs of between six months and a year, titles like Fellini's *La Dolce Vita*. The more I would read in *The New York Times* about all of the interesting foreign films playing in New York, I wondered why we weren't seeing them in Chicago.

We were missing a world of films because the theater owners wouldn't take chances with different unknown foreign movies from anyone but the masters. They stuck only with a guaranteed winner like a Bergman or a Fellini, forever.

I was making short films by then, blew one up to 35mm and started entering in foreign film festivals myself. If I couldn't bring international cinema to Chicago, I'd bring Chicago to international cinema. My little film *Eve: Story of a Lonely Girl* played in Cork, Ireland and Krakow, Poland, at their short film festivals. It was the tale of a lonely little girl who befriends a dog. Today, that same girl would have to get raped or shot or hit by a car. A much more innocent time.

I asked one of the art house theater owners if my short could play in Chicago before the feature film. One said yes, and it did. This left me determined to find a way to show my little movie in more places.

My next short received a special Diploma of Honor in Cannes, France, at a festival. Ceci and Jim O'Riley, a couple I had met along the way in Chicago, along with Hubert Schmuttenmaer (and trust me, you can't make up a name like that) agreed with me about what we were missing in Chicago. Jim was an editor of industrial films. Ceci was a minor socialite who knew people.

The three of us got a group together and kicked around the idea of starting a large film event that could show stuff like my own work and honor the filmmakers, student films, documentaries, you name it. It was important to award prizes and introduce new, first-time directors while having the films judged by their filmmaker peers. The idea was that everyone needed a slap on the back and someone telling them, "Hey, you're special."

I convinced my father to underwrite the festival with the understanding that it would not stop me from pre-med at Loyola University – and that this thing was "just a phase" on the way to my becoming a doctor. I'm sure the fact it never happened broke his heart. My mother's, too.

The festival idea quickly took center stage in my life and never left, and these friends all volunteered to make it happen. I was on and off dating Judy Kohnke from Loyola. She graduated and worked at a design firm but did all the hard work and correspondence needed to start the festival.

Starting the festival in this magical city was easy. People wanted to enter their films and were drawn to it. We sent out this mass mailing of like 10,000 pieces to every film company and institute in the world. The name "Chicago" resonated, as it represents something magical because of Al Capone and The Legend. The town had a mythic quality that drew interest from the get-go.

It was actually getting an audience that proved the tough part.

I scheduled a showing of my own little film at the Ambassador West Hotel with drinks and some food. I invited the press, and they actually came. It was there that I initially introduced the idea of an international film festival. It was also there at this little event that powerful *Chicago Sun-Times* columnist Irv Kupcinet took a liking to me and the cause.

Soon enough, an introduction he made would change my life.

2 – A STAR IS BORN

Irv Kupcinet was more than just a newspaper columnist. He was a legend, sort of the Walter Winchell of Chicago. (If you don't know who Walter Winchell is, Google him.)

They called Irv Mister Chicago, since he carried so much power and weight in all aspects of things important to his readers: movies, theatre, celebrities, sports, and politics. Kupcinet had his finger on the pulse of my city. He had heard about me, this kid from the West Side of Chicago who made little films and wanted to start a great big international film event right here in our fair city.

Irv took me to lunch one day at Fritzel's Restaurant, where he hung out. It was the place to see and be seen. He told me, "I'm going to introduce you to someone who might be able to help you. She was a famous silent film star in the '20s and early '30s and retired to Chicago. Her husband has recently died, and I think it's time for her to get over the grieving widow thing. You just might be what she needs right now. Her name is Colleen Moore Hargrave."

That's how it began. Colleen was a socialite and a widow, and Irv felt it was time she got back into the movie game. That's where I came in.

I made an appointment and met with Mrs. Hargrave. She invited me over to her place for lunch. I put on a suit and tie and headed to 1320 North State Parkway. Greeting me there at the door was a real-life Auntie Mame.

Now you remember who Auntie Mame was, right? The Patrick Dennis book, play, and film about an irresistibly dynamic widow. That's who Colleen was to me. She would ultimately become my second mother.

Mind you, I had neglected to do my homework on just whom I was visiting. I remembered that Kupcinet had said she was a silent film star. But that was about it. When entering her fabulous apartment, I noticed over the fireplace sat a giant painting of a young woman. I assumed it was Colleen when she was a movie star. The painting was probably five feet high, depicting a cute gal with a short haircut. I later learned that her haircut was synonymous with her character in the films.

She and I hit it off immediately. I'm not sure why. It may have been my sense of humor. She would always say, "I need a good laugh." I was good at that, at searching to find a lighter side to most everything. Growing up with two doctors makes you try to uncover some fun around the corner.

I was back at Colleen's the next day for lunch. This time, another person was sitting at the table with us. It was one of her best friends who was visiting, a legendary actress named Myrna Loy. Once again, brilliant me had done no homework, and I was sitting beside someone whom I'd seen and loved in movies but didn't know it.

What I came to realize was that many of Colleen's old movie star friends were coming to visit her in Chicago to see how she was doing. They would stay across the street at the famed Ambassador East Hotel with its celebrity restaurant The Pump Room.

Colleen loved the idea of an international film event. I explained to her what festivals were, how we could make this one really big for Chicago, and that she would be the honorary chair. It was instantly clear to me that she needed something new in her life, and she felt I was it. Smart woman, I say.

Mind you, it wasn't a no-brainer to put on a Chicago festival back then as much as you'd think. It wasn't like today where every town had one. There was no Sundance yet, no Toronto, no Montreal. It was Venice that really kick-started the idea of international film gatherings way back in 1932, when Mussolini (of all people) founded it as a way to rustle up positive publicity and promote tourism. Cannes didn't come into existence for another 14 years.

But anyway, Colleen took on the challenge with gusto. She started inviting her lady friends from Chicago to meet me, and I realized they were not just people in her circle but the most powerful women in the city because they were married to the most powerful men in the city. I remember one lady telling me, "Oh, don't try to remember my name. Just remember that I'm the electric company." Another said, "Oh, I'm the Art Institute." It was really kinda hilarious.

I recall that these women all smoked and loved to drink. They were saucy, smart, independent gals – and they loved me. I guess they figured I was cute and relatively harmless. Plus, I had Colleen's stamp of approval.

Colleen finally said, "It's time to introduce you to Chicago business society. We'll need to get them involved." We went to lunch at a place called the

Arts Club, where the powerful people would meet to discuss powerful things in a powerful way. I think I was the youngest person at every lunch by a wide margin.

Our hostess would tablehop, since she knew all these people thanks to her late husband Homer Hargrave, who was a founding partner of Merrill Lynch, the investment company. I wasn't very good at remembering names, but all of these elderly gentlemen represented major corporations in Chicago, and Colleen felt they should be involved with this new film festival brainchild.

I think I gained twenty pounds at this series of Arts Club luncheons with all of the heavy food and a really awful salad dressing that was evidently famous, I guess, as they even sold the stuff there. Contacts were being made, boards were being developed, and the idea of the festival was becoming very real. Momentum was building, and quickly.

Back at Colleen's apartment, she said, "Tomorrow we're having a special guest for lunch. Her name is Joan Crawford." Once again, I knew she was a famous Hollywood movie star, but I didn't know much more than that about her.

How was Joan? She was angry, and very businesslike. She also looked remarkably old but well put together. I didn't know a thing about her, and I wasn't smart enough to do research that honestly would have helped me a lot.

Yeah, I was really pretty stupid in those days, being so young and not realizing what I had at my fingertips. It was about this time that Colleen said, "You know, you're too young for this role, so we're going to make you 27 years old."

She gave me a pair of glasses. I think they were from her old fellow silent film star Harold Lloyd. She ordered, "Wear these glasses, and from this day on you will be 27 years old. (I was 22 at the time.) No one trusts someone who's 22."

Of course, she was correct, and I delighted at this transformation.

Colleen mentioned that Crawford could only stay for a short time. She was here for the Pepsi Cola convention. as she was on the company board and warned me that her makeup had "only a few hours" of staying power. I had no idea what that meant. Colleen explained that Joan placed gauze pieces in various places on her face to lift areas (known as "wings"), and then placed the makeup on top. It worked wonders until it didn't.

I was learning. Slowly.

Colleen did have a very specific (and very aristocratic) way of doing things. Example: I became good friends with her assistant, a lady named Annie. She did all the secretarial work and helped Colleen serve dinners and things. I mean, she did everything for her.

One evening, Colleen invited me to one of her dinner parties with lots of cool people. Annie was serving the food. Anyway, I needed something and approached her with, "Annie…"

Immediately Colleen shut me up, basically saying, "We don't speak to the help."

"But it's Annie," I reasoned.

Didn't matter. Help was help as far as Colleen was concerned. I never made that mistake again. It was like she became a bit of a different person to me from that moment on.

As I grew older and (somewhat) wiser with Colleen, I started to read up on who I was to visit with next so I didn't embarrass myself by not knowing a big star. This one happened to be Lillian Gish, for lunch at The Pump Room across the street. After that, director King Vidor came to visit, and I learned that he helped discover Colleen in a film called *The Sky Pilot*, after D.W. Griffith saw her screen test and hired her for $50 a week in L.A.

Colleen seemed to be having this on-again, off-again affair with King. Of course, King was still married back in Hollywood, but that didn't seem to matter, as Colleen had a few husbands along the way herself.

Back at the Arts Club while table-hopping, we came upon two major advertising moguls, Leo Burnett and Fairfax Cone (yes, that's his real name). Both were heads of major ad agencies and sitting separately. of course. Colleen knew each of them, and now I did too. They agreed to be on the board if I would include television as a category in this new festival. Burnett said, "One of my sixty-second television commercials says more than any of those fancy foreign films of yours, and in less time." He was right, actually.

I was always a big fan of TV and commercials from that moment on. Telling a story from beginning to end of half a minute or a minute remains a phenomenal skill.

Very few are aware that both *What Price Hollywood?* (1932) and the original *A Star Is Born* (1937) are based on the early life of Colleen Moore and her marriage to John McCormick, a film producer and publicist as well as a suicidal alcoholic. Journalist and screenwriter Adela Rogers St. Johns wrote both movies and was one of Colleen's closest friends throughout her life. She even ghosted most of Colleen's autobiography *Silent Star.*

Colleen's rise to stardom reads like a fairytale. It was the story of a little girl who could act and wanted only one thing: to be in the movies, From her screen test in Chicago at the Essanay Studios to having that test shown to Griffith as a favor owed, it was a story of kismet.

She was signed for six months at the age of 14 and was able to perform as a serious young actress in as many as four short films a year. Colleen moved from studio to studio and sometimes acted in two films at the same time. She was a teenage businesswoman, increasing her salary and fame along the way.

Mind you, Colleen's roles were always the same. She was either the girl-next-door type, or she was kidnapped, or maybe she got the guy in the end (or not at all). She was the naïve innocent. Meeting Vidor in 1921 changed her life and pushed her into features. Yet she was being continually type-cast, and she knew it.

Colleen saw, for example, that Gloria Swanson (also at Essanay) was growing tired of comedy and the whole Keystone Kops thing. She wanted to do drama. Colleen was the opposite. She was done with drama and thought she wanted to do comedy. So, she began selling herself to the various comedy studios in Hollywood. She wasn't very good because she lacked the requisite comic timing, so she played the straight woman to the foolishness unfolding around her.

The 1920s and the age of the flapper on the screen was the next step for Colleen. She knew she wasn't the sex-drugs-booze kind of actress, so she made do as a clean-cut flapper! That made her totally different from the rest, but it worked. Her husband McCormick had the film script that would make her a star, *Flaming Youth*, followed by *Why Be Good?, Her Wild Oat, Synthetic Sin,* and the biggest one, *Lilac Time.*

Colleen was the highest paid actress in the country between 1927 and '29. She successfully moved from silents to talkies in *Footlights and Fools*

('29). She sang, she danced, and she even affected French and Irish accents. The woman was a workaholic and at the same time terribly unhappy with her personal life; she really wanted out.

With her continued success in films, her hubby's drinking escalated. It was in New York at a film promotion that John attempted to kill her. Colleen had faced his drunken rage first in 1927 when she asked for a divorce. After fuming, "You can't divorce me! You're Catholic!", McCormick attempted to push out of the 16th floor balcony of the Ambassador Hotel in New York. Colleen was saved by her best friend, the soon-to-be-great director Mervyn LeRoy, who fortunately also happened to be in the room.

Yet Colleen forgave John and put him into alcohol treatment. After *Footlights and Fools* opened, he once again attempted to kill her at their multi-million-dollar Bel Air mansion. This time, the waiting chauffer saved her life. Colleen finally called a lawyer and filed for divorce after years of abuse and close calls. What a shitstorm.

Why did Colleen stick around so long? Well, she felt that her husband had created her, so she owed him. But he didn't have a right to completely control her life or try to end it. Unfortunately, as an abused spouse, she bought into his ownership of her and stayed with him come hell or high water – and both came.

But now the marriage was over. Unfortunately, so was Colleen's contract at the studio.

Colleen kept the Bel Air house and gave John their Malibu beach house, once again feeling guilty about the whole mess. Colleen went on to a few more serious sound feature films in 1933 and '34. One (*The Power and the Glory* with Spencer Tracy) was successful. The other (*The Scarlet Letter*) was not.

At this point, Colleen decided to leave the world of film for good and remarry. Her new husband was a fun-loving guy called Albert Parker Scott. He had a seat on the New York Stock Exchange, and for the first time Colleen was joyous, laughing, and having fun. Albert pulled together all her finances for her and helped her realize she was a very rich woman.

I guess it was during their honeymoon at Niagara Falls that Colleen discovered that Albert was playing for the other team, as they say. They immediately separated and remained close friends, since Colleen always needed a good laugh, as she often said. During this time, John McCormick actually did walk into the Pacific Ocean to commit suicide, as in *A Star Is Born*, but was saved by his next-door neighbor, actor John Gilbert. He would remarry a few more times and live into his sixties.

Having left her film career behind, Colleen devoted herself to The Colleen Moore Doll House, an elaborate fairy castle housing more than 1,500 miniatures that in the mid-1930s carried an astonishing price tag of nearly $500,000 (or more than $10 million today). This kept her name and identity alive beyond cinema. She also traveled the United States extensively.

Colleen would marry for a third time, in 1937 to Homer Hargrave, leading to her becoming a major player in Chicago society. Homer gave her a ready-made family with children that she adored as her own. When Homer died in 1964, she had the Doll House, her family, and then she met me and backed the film festival as a new adventure that would bring many of her film friends from decades before back into her life.

It always saddened Colleen that so few of her films survived. Out of her 49 movies, maybe five remain in circulation. She claimed to have given the entire collection of First National productions to the Museum of Modern Art in New York City, yet MOMA claims to have no such collection. This mystified and frustrated her.

From that moment on, Colleen made a valiant attempt to find her lost films. Whenever she traveled country to country, she would inquire at film institutes and museums. Some found a few fragments here and there. Eventually, a couple films were successfully restored.

I never spoke to Colleen about her personal life. Of course, I knew she had continued her close friendship with King Vidor, more as an escort than a lover. She'd say, "He's just so old for me." Suddenly in 1965, along came a new man in her life, a French doctor who was a big deal at the World Health Organization in New York. Colleen seemed smitten with him.

With her new infatuation came Joel, the doctor's 21-year-old son that Colleen wanted to help by bringing him to Chicago, having his teeth fixed, and using him as her driver. Oh yes, he was a filmmaker or film producer, she was never clear or sure which, but she assured me, "He can help you with the new film festival we are starting."

Joel's ability with English was fair, and he seemed harmless enough. After a few accidents, however, Colleen didn't want to use him as her driver any longer. It turned out he really wasn't a filmmaker, either. Big shock. Nonetheless, she was stuck with him. And so was I.

"Michael, Joel needs a place to stay," she said to me, "so you take him in and he'll help with the festival."

"But I have no place for him to sleep," I replied. "All I have is a single twin bed in my apartment."

"So, he can sleep with you!" Colleen concluded, thinking herself a genius.

The first thing I noticed was that Joel always smoked these really strong French cigarettes and alternated them with marijuana, so he was continually high. I've decided that all French men are bisexual, and here I had the best proof right in my bed with me.

I called Colleen the next day to explain that I didn't think this was going to work out.

"Michael, you have to try everything in life," Colleen counseled. She convinced me that Joel would be my boyfriend, partner, lover, whatever. I found it pretty exciting, as he was always high and really got into sex all the time. I guess I was falling in love with the guy.

Joel was useless at the film festival, however, so I sent him off and he spent a lot of time doing odd jobs for Colleen and her family. We did dinners together. He was a terrific cook. I mean, c'mon, he was French. It remained a sexual thing until he would disappear for days at a time. It turns out he was having a hot and heavy sexual relationship with a cute little French girl at the same time. I was crushed, of course, but we all became close friends, though that didn't mean any *menage a trois* liaisons.

Next thing I knew, Joel and I were traveling together. Colleen went to the Cannes Film Festival with Vidor, who was on the jury. I went scouting films, and Joel tagged along as my assistant. He was useful with his French, and he turned out to be a rather good photographer. He was so charming that he could crash any fabulous party, so he met lots of the young actors and actresses and did get sexually involved with a gay French director whom he would get high with on the beach. Do you sense a pattern?

Joel was on his own and in a good place. He was also playing around with other drugs by then. After Cannes, he went back to New York to be with his father and I never heard from him again until one day when I was summoned to my father's office. Now remember, my dad was a doctor as well as the president of St. Mary's Hospital here in town. I was terrified.

Dad confronted me with a medical alert from the New York Board of Health with my name on it, for being exposed to syphilis, gonorrhea, you name it. Joel was obviously infected, and as he went to a public hospital, he had to disclose the names of everyone he was involved with sexually.

I don't recall whether the report mentioned if it was a male or female who had infected him. My dad didn't ask me either, thank God. I never did tell Colleen any of this, but thinking back, she might have gotten a good laugh out of it.

3 – THE GREATEST SHOW ON EARTH

Every film festival has humble roots.

Consider the Sundance Film Festival. It began life in 1978 as the United States Film Festival before it was purchased by Robert Redford and rebranded as Sundance, to be paired with the film institute of the same name that he was building in a Utah mountain retreat. It would take years before it became the big deal it is today.

In case it isn't already apparent, I got there long before Redford did. Just making that clear.

Our little festival began on November 6, 1965, a cold and snowy night in Chicago. The movie theater was in the Gold Coast area, which is a rather ritzy part of Chi-town. It was at the Carnegie Theatre on Rush Street, a small art house that was popular in those days for showing foreign films (as all art houses did).

We had one spotlight shooting up into the sky and a big black limousine waiting in front. It looked like it was the opening night of a big deal, and it was: the kickoff of the first Chicago International Film Festival.

The only problem was, no one showed up.

The producer, director, and stars of the film kept asking, "Where are the people?" I had no answer. All I knew was, they weren't there. And it was more than a bit embarrassing.

I had chosen a small independent American comedy entitled *Saturday Night in Apple Valley* that was getting a lot of play on late night TV talk shows as our opening night film. It was a world premiere, so that sounded important. It had a few small stars, Phil Ford and Mimi Hines.

But again, nobody showed. And I mean NOBODY. I was staring out at 500 empty seats.

I thought we had everything covered. There were seven competitive categories of films and screenings at four college campuses showing the movies free of charge. We had advance positive press. All four newspapers in town announced the first festival as an exciting idea and one long overdue for the city. We even got publicity from none other than Bette Davis, who was in town promoting a new film of her own.

We'd scheduled a powerful lineup of heavyweights coming to the festival, including a pair of Academy Award®-winning directors, Stanley Kramer and King Vidor, as well as Italian directing great Federico Fellini. What more could you ask for?

We welcomed 300 entries in total, including features, shorts, documentaries, student films, even TV commercials. Our commercial category was unique in that home audiences could watch it "live" on TV and call in their votes for the best spot. Remember, this was 1965.

I figured that all of this advance visibility would guarantee a festival audience. News flash: it didn't. The few foreign films presented didn't move critics or the public. Two excellent documentaries, one by William Friedkin and the other by Kristoff Zanussi, didn't make any waves, either. The directors' names had yet to become known.

On the positive side, the Stanley Kramer masterclass was a big success. It was held at Mundelein College, a Catholic school on the campus of Loyola University, and he was terrific. The clip reel of films he either produced or directed was excellent. It featured scenes from *Inherit the Wind, It's a Mad, Mad, Mad, Mad World, Death of a Salesman*, and *The Wild One*. All the nuns lived in a convent at the school, and they had top-notch equipment (35 millimeter!) because they never left.

Meanwhile, our Italian star director Mr. Fellini didn't fly in, claiming he was bedridden in Rome. Oy.

Our closing night was moved to the new Playboy Theatre. Yes, *Playboy* magazine had its own theater. Thanks, Hef. We held our "Best of the Festival" event there, showing all the winning films in a row. Those screenings *did* manage to sell out, proving to me that the public had heard about the festival and just wanted to see the best stuff all at once.

Publicly the first festival ended that night, but privately the festival continued with an awards night gala. It was a fancy dress ball, by invitation only. This was the project that my chairwoman Colleen Moore had been slaving over with her team of powerful, WASP-y rich ladies. What does WASP stand for? Why, White Anglo-Saxon Protestant, the upper crust of American society.

Colleen became one of the ritzy through her marriage to Homer Hargrave, and she had rounded up the most dazzling society dames in Chicago to put this event together. Keep in mind that she was a famed onetime silent film star, so we held the gala on the very same soundstage of the Essanay Studios where Colleen had performed her first screen test in 1920 before going to Hollywood.

Essanay was the equivalent of Hollywood in Chicago. You had Gloria Swanson working there as well as Wallace Beery, the Keystone Kops, and even Charlie Chaplin made one short film there.

So, imagine five hundred white ladies and gentlemen dressed to the nines at an awards gala on a soundstage covered in white linens, white balloons, and giant stills from King Vidor's films, projected all over the giant floor-to-ceiling, white cyclorama walls. It was a spectacular sight.

King was a legendary director who bridged the eras of both silent and sound films from *The Crowd* (1928) and *The Big Parade* (1925) to *The Fountainhead* (1949), *Duel In The Sun* (1946) and *War and Peace* (1956). The stage presentation was touchingly given by Colleen to King.

How did we get King Vidor to even show up in Chicago? Well, it didn't hurt that King and Colleen remained very close, sometimes close enough to turn over and say hello to each other in the morning. However, there was a moment at the start of the gala as guests were checking in to the receiving line. Colleen pulled me aside.

"There's a colored woman coming in," she said by way of alert.

Some of Colleen's ladies had warned, "We can't have a colored person at our party."

I honestly couldn't believe it. I was appalled, as I had never seen this side of Colleen and her ladies. Keep in mind that I'm just a 22-year-old kid faced with my chairwoman going ballistic. Fortunately, Irv Kupcinet was coming in just behind the black lady.

"Kup, help me, I have a situation," I pleaded.

Irv quickly assessed what was going on and introduced the black lady to Colleen.

"Colleen, I would like to present to you a Pulitzer Prize-winning author, Miss Gwendolyn Brooks. She is the Poet Laureate of Illinois and is representing the mayor of our city this evening."

Colleen welcomed Gwendolyn with open arms. Thank God for Kup. It wouldn't be the only time he saved the day. The gala was a smashing success for King, Colleen, and the guests, but it was *thisclose* to being a complete disaster. One racist incident could have ruined the whole thing.

The next day, the film festival reviews arrived.

"Chaos, too large an undertaking," proclaimed one. "Kutza took on more than he could handle and complete indifference from the city where it was held are the factors that contributed to the festival's gloomy first try. There were some bright spots but too many loose ends. The best thing was the 'Best of the Fest' event and Colleen Moore's covering up of the holes with her motion picture and society relationships, giving glitter to the event."

All in all, ouch!

I was heartbroken with the wrap-up coverage. Colleen called me and reassured me, "Michael, nobody cares about newspaper critics and reviews. They wrap their dead fish in the papers and throw them out the next day."

I was hoping she was right. About nobody caring about critics, I mean, rather than the dead fish.

--

Flash forward: it was the next year and the closing of the second Chicago Film Festival, during the afterparty at the famed Playboy Mansion hosted by Hugh Hefner himself. It was a wild, elegantly staged event with endless food, drink, music, and young vibrant guests dancing and schmoozing.

Colleen and the legendary silent actor/director Harold Lloyd sat in one corner trying to stay awake and enjoy the party. The next day, a newspaper critic commented on these "two old, once famous silent movie stars falling asleep in the corner wondering what they were doing at such a party past their bedtime?" Talk about classless.

Anyway, Colleen rang me in the morning screaming, "How can they say these things about me?"

"Colleen," I replied, "nobody reads newspaper critics. Remember? They just wrap their dead fish in the papers and throw them away the next day."

"But this is different!" she insisted. "It's about ME!"

And there you have it.

The good news was that I seemed to make up in charm and humor what I lacked in understanding and business savvy. It's really the secret to my success, a badge I wear proudly to this day. At the same time, I did it without being a star fucker. I never tried to hang with celebrities or make them my best friend. I just wanted them to knock everyone's socks off at the festival.

For the most part, I succeeded in that regularly. I helped to create something special from nothing. The Chicago International Film Festival would grow to become a big-huge deal around the world — but ironically, significantly less so in my own city.

I recall a famous line that proved consistently and undeniably true: Chicago eats its own children. I would find out just how accurate that was in the years to come.

As for Colleen, she ultimately left Chicago and was unofficially linked with King Vidor until King's death in November 1982. The following year, she married her fourth and final husband, a builder named Paul Magenot, in 1983. He built Colleen her dream home in Templeton. It was located near the Central California community of Paso Robles where Vidor had his ranch and not far from the famed Hearst Castle in San Simeon, the home of publishing mogul William Randolph Hearst. It's a place where Colleen enjoyed so many lavish celebrations back in the day. That was during the time when Hearst was having an illicit sexual relationship with her fellow silent film star and dear friend Marion Davies.

The area carried so many memories for Colleen. Unfortunately, she wouldn't live long enough to enjoy that new ranch spread very much. She died on January 25, 1988, at either 85, 87, or 88 years old, depending on who you believe.

Yes, Colleen kept 'em guessing to the very end — and beyond. And I loved her for it.

(clockwise from upper left): My father Dr. Michael Kutza Sr,
my brother Peter, my mother Dr. Theresa Felicetti,
and little old me. (Michael Kutza Archives)

My late grandmother in her coffin. She lay in our living room for a full week, leaving me forever traumatized. (Michael Kutza Archives)

The radiant silent film star Colleen Moore on the set of the 1929 movie *Synthetic Sin*. (First National Pictures)

Young me surrounded by silent movie greats – Lillian Gish on my left, Colleen Moore
on my right – at the Pump Room in Chicago in 1965. (Chicago Festival Staff)

Colleen Moore and the great filmmaker King Vidor at the Chicago
International Film Festival. (Chicago Festival Staff)

My incomparable friend and fundraiser Mary Ann Josh is standing at left, with Colleen Moore seated at right. (Chicago Festival Staff)

That's me seated beside the marquee prepping for our first Chicago Festival in '65. (Chicago Festival Staff)

Colleen Moore stands in front of a painting of
her younger self in 1966. (Stan Lazan)

Speaking at our first Chicago International Film Festival Gala.
That's columnist Irv Kupcinet on the right and a pic of silent film
legend Harold Lloyd on the left. (Chicago Festival Staff)

In '65 at our Chicago Festival Gala. Yes, it's a
little WASP-y. (Chicago Festival Staff)

With the great Angela Lansbury, whom the
festival honored in 1974. (Chicago Festival Staff)

Hanging out with the kids from our children's film jury at an
early Chicago Fest. (Chicago Festival Staff)

Irv Kupcinet (a.k.a. "Kup") and his wife Essee. (Kupcinet Family)

The great Italian filmmaker Franco Zeffirelli (right) and
Playboy magazine's West Coast Photo Editor Marilyn Grabowski,
a great friend of our festival. (Chicago Festival Staff)

Film critic Roger Ebert in his youth. We helped put
each other on the map. (Chicago Festival Staff)

(left to right): Chicago Festival board member Carolyn Leopold, me, Chicago Mayor Richard J. Daley, and our executive director Lois Stransky before our 10th festival. (Chicago Festival Staff)

Ballerina and choreographer Ruth Page (in hat) and the great French filmmaker Louis Malle hang at the festival. (Chicago Festival Staff)

(left to right) Superagent Michael Ovitz, Mr. and Mrs. Sydney Pollock, and Oliver Stone's mother Jacqueline. (Robert Dowey)

With film producer Arthur Cohn (left) at the Manila International Film Festival in 1972. (Manila Film Festival)

Me pitching Orson Welles to come to my fledgling festival in
Chicago at the 1966 Cannes Festival. (Cannes Film Festival)

French filmmaker Agnes Varda with Roger Ebert
at Cannes. (Cannes Film Festival)

Tom Cruise, after receiving the Chicago Festival's award for
Actor of the Decade, at our Gala with his mother Mary Lee Pfeiffer
and stepfather Jack South. (Chicago Festival Staff)

Sally Field and her young sons at our Chicago
Festival Gala. (Robert Dowey)

The sexploitation filmmaker Russ Meyer. (Robert Dowey)

The great filmmaker Robert Altman reaches for some popcorn during a screening at our 1982 Chicago Festival. (Chicago Festival Staff.)

Our Chicago International Film Festival featured the world premiere
of the future Best Picture Oscar® winner *One Flew Over
the Cuckoo's Nest* in 1975. (Chicago Festival Staff)

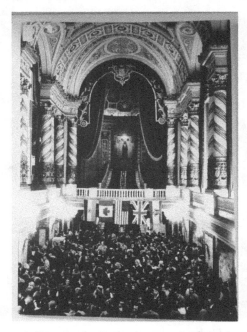

The 3,400-seat Granada Theatre on the premiere night of
Cuckoo's Nest. (Chicago Festival Staff)

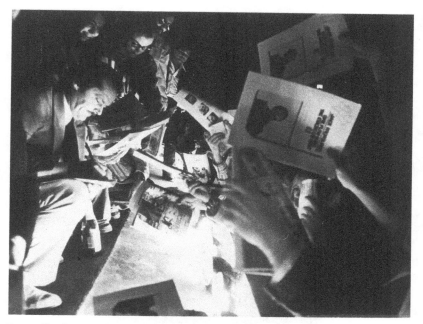

Oscar® winner Jack Nicholson is besieged by fans following our
Cuckoo's Nest premiere. (Chicago Festival Staff)

Jack Nicholson reacts to the buzz on stage following our
Cuckoo's Nest screening. (Chicago Festival Staff)

Cuckoo's Nest Oscar® winner Louise Fletcher at the
Chicago Festival in '75. (Chicago Festival Staff)

The Oscar®-winning *Cuckoo's Nest* director
Milos Forman. (Chicago Festival Staff)

Actress Karen Black (left) with film critic and author
John Russell Taylor in 1979. (Chicago Festival Staff)

French director Jacques Demy. (Robert Dowey)

The incomparable Mel Brooks making his usual entrance
at our Chicago Festival. (Chicago Festival Staff)

The legendary eight-time Academy Award®-winning costume designer
Edith Head (second from right). (Chicago Festival Staff)

(left to right) The Oscar®-winning director William Friedkin and film critic Gene Siskel. (Chicago Festival Staff)

Ray Nordstrand (left) and Charles Benton, both Chicago Festival chairmen at different times. (Chicago Festival Staff)

The author in Hawaii in the 1970s in all his naked glory. (Nicholas Cann)

Albert Johnson, the San Francisco International Film Festival program director and a dear friend to my Chicago Festival, sharing a laugh with me in Manila. (Manila International Film Festival)

British director Mike Leigh. (Chicago Festival Staff)

Italian actor Marcello Mastroianni (left) and Playboy's West Coast
Photo Editor Marilyn Grabowski. (Chicago Festival Staff)

4 – SEX AND THE CITY

It was 1966, and I was watching some TV show comprised of short little films made by Chicagoans. Suddenly, there came one by a famous local fashion photographer. It was titled, *No Comment* and was in black and white, featuring a bunch of flashy photos of pretty people backed by music, all of it intercut in a hypnotic way that caught my attention.

I then read in the next day's *Chicago Tribune* about this same photographer, Victor Skrebneski, who was featured in the short, praising a new advertising campaign of his. I thought that his little film should be in my festival, and secondly, I'd like to be in that world of those fabulous people somewhere in Chicago's underground nightlife.

I searched out his studio address and decided to pay a visit. What I found was like an armed encampment controlled by two strong women, one tall, one short, a barking little awful rat-dog, and some unpleasant partner who seemed angry and drunk.

Since I could see this wasn't going to be easy, I started with the tall one, Jovanna. She was Greek, and her middle name had to be "No," because that's all I could ever get out of her.

"Mr. Skrebneski is busy," she'd say dismissively. "He's with a client. He isn't interested. Go away."

Years later, I understood the whole deal. Every model in town would come in and leave a photo composite, and Jovanna would throw them out.

Now the short one, Ida, was Jewish and not nearly so rude. She asked, "You're the one in the press lately with the film festival and Colleen Moore?"

"Yes," I replied, "and I would like to show Mr. Skrebneski's film."

"Let's have lunch next week," Ida suggested.

The rat-dog kept barking at me, and the boyfriend or whatever kept glaring in the background, as if to say, "Keep away from my property!"

That lunch with Ida was a turning point in my life. First, I learned that she was once an actress and was in the 1926 Broadway show entitled *Sex* starring Mae West, where they all famously got arrested and taken to jail in New York City.

I loved her already.

I told Ida all about the film festival, which had effectively flopped a few weeks earlier (except for the star-studded gala closing of course, with every powerhouse couple in Chicago attending). Ida knew I could be useful to Victor. I told her about Cannes and all the naked stars jumping into fountains to get noticed, and how Chicago's festival needed to become sexy if it was to have a second year. And I made sure she knew that I wanted Mr. Skrebneski to help make that happen.

Ida set up the meeting with Victor. It was a business meeting, but friendly, and he said, "Give me a few weeks." That was it. Mr. Skrebneski, now Victor, helped make us instantly famous with a single photo. It depicted a naked model named Paulette Lindberg, soaking wet, holding our award statue in her arms. It was very phallic, but I didn't realize that at the time. (I was still pretty naive back then.)

"This is hot!" I said. I was thinking we'd run it as a full page in *Variety*. We bought the ads, and the text shamelessly announced, "Come to Chicago and win your award." It was instantly noticed because of course the model was dripping and beautiful and shadowy sexy. And... It instantly put us on the map. The world was taking note of the Chicago International Film Festival because of Skrebneski's photo. That's all it took.

Victor would come up with years of outrageous wet t-shirt posters. We had to stop using the picture of Paulette because feminists had just become very volatile and came after us with both barrels. We were being ostracized by groups screaming, "How dare you exploit women like this!"

"Are we exploiting women? This is fantastic," I replied. I didn't consider it a very serious problem. It was easy enough to throw in a few naked men to feel exploited and even out the score. And anyway, this was Chicago, the home of *Playboy* mag. Give me a break. Sheesh.

One of the most sought-after film festival posters that Skrebneski ever shot was one featuring the ill-fated Anna Nicole Smith. Surely you know who Anna Nicole is. She started out as the Guess Jeans model and moved up to become *Playboy's* Playmate of the Year as the prototypical big, beautiful woman. She was a larger-than-life blonde who wanted only one thing in life: to be the next Marilyn Monroe.

MICHAEL KUTZA

My first meeting with Anna took place at the airport after picking her up. We had invited her to Chicago to be on the next poster. The photo editor of *Playboy*, my friend Marilyn Grabowski – who also introduced us to Sharon Stone – suggested Anna.

The young lady I met was a twentysomething country girl straight off the farm, or so it seemed to me, I didn't know Anna had been a topless dancer in a few clubs before all this modeling. She had a lot of cash on her, as she had just completed her role as the Guess Jeans model, and planned to hit the swankiest shopping mall in Chicago before checking into the hotel.

I had our limo driver head over to 900 North Michigan Avenue, and Anna had no trouble finding the expensive shops. She had to buy gifts for all her friends back home with all her dough, because God forbid she should save any of it. Her biggest discovery was Charles Jourdan shoes. A $2,000 pair caught her eye, and she simply had to have them. This all took about an hour before we finally headed to the hotel.

No sooner had we hit the Swissotel in downtown Chicago than Anna marched right up to the concierge and said, "Please FedEx all of these gifts to my address in Texas." Bam! Then she checked in and headed to the penthouse suite, which was generously donated to us by the hotel, represented by my dear friend, Robert Allegrini.

Robert did the walk-thru with me and Anna, room after room in the massive suite. The fireplace is what seemed to fascinate Anna the most. It was gas and looked remarkably like a natural woodburning one, what with its logs burning. Once she found the remote for the fireplace, she was content to just play with it like a little kid.

Anna's luggage arrived, and I left with Robert, scheduling to return later to pick her up for the photo shoot and dinner afterward. After we departed, Robert realized he still had the suite key in his pocket and went back up to return it. She didn't answer the door, so he let himself in to leave it on the table. Anna came out clad only in a skimpy towel to thank Robert. To this day, he tells the story of the near naked blonde vision thanking him.

The shot that Skrebneski came up with out of the shoot was amazing as always. Anna met our male model Mark, and the two of them hit it off beautifully, as you might expect, since they were nude except for the film festival t-shirt covering up a bit. Anna was clutching the shirt, holding her breasts up while pressed against Mark. Ooh la la!

Later, we all went to dinner at a nearby chic restaurant. Anna was a riot, requesting ketchup for her mashed potatoes, since they didn't serve french fries. This place didn't seem to even know what ketchup was. But that was nothing compared to her next request.

"I need rubbers!" Anna declared.

Yes, now that she was totally hot for Mark the male model, Anna expressed her intention to fuck him. Which she did, of course. The poster turned out to be a sensation, so it was all more than worth it.

Anna would go on to enjoy a remarkable couple of years. There were a few bit parts as a blonde bombshell in *The Naked Gun* and *The Hudsucker Proxy*, a marriage to an 89-year-old oil tycoon, a breakout hit *E!* reality show, and the media seemingly following her every move. Then success and drugs got in the way to take her down.

There would be no happy ending for Anna Nicole Smith. But I'll never forget meeting what seemed like a wide-eyed gal from the sticks. She mentioned to me from the start that we would like each other, as we shared the same birthday, November 28. And she was right.

--

In all my years as the artistic director of the Chicago Festival, I was always on the lookout for sensitive sexual subjects that hadn't been seen much on the film festival circuit.

As far back as 1969, I'd come upon a film or two with subtle gay themes before they became so trendy. We would highlight the work of German master Rainer Werner Fassbinder, Israeli director Amos Guttman, Mexican director Jaime Humberto Hermosillo, French directors Patrice Chereau and Andre Techine, and American filmmakers John Cameron Mitchell, Gus Van Sant, and Bill Condon.

I saw that there was a genuine niche audience developing for these films and even created a special section called "Out-Look" complete with a jury and awards. LGBTQ became a big deal. The themes ranged from hard-hitting dramas to lesbian romances to documentaries about famous gay icons, all superbly crafted perspectives on sexuality and identity from around the world.

I came upon a Chinese-born filmmaker who worked mostly in Hong Kong at the Berlin Film Festival in 2010. His approach was stylish and erotic, focused mostly on ravishing young men. It wasn't quite porn, but almost. His work was winning awards, and I thought it would definitely shake things up in Chicago at our festival.

His name was Danny Cheng Wan-Cheung, but he went by the nickname Scud. I have no idea why, so I asked. He replied, "It translates to my Chinese name 'Scudding Clouds' in English." He's a sensation in his incorporation of taboo themes in erotic same-sex cinema.

Scud was inspired by Pier Paolo Pasolini and Pedro Almodovar, and it shows in all eight of his feature films. What I like about the guy is his love of pushing boundaries. His films aren't just pretty naked boys but some hot women, too. His subject matter can get pretty gritty and realistic, to the point of death, dismemberment, and rebirth. Heavy stuff.

I recall how critic Gene Siskel wrote, "All Kutza does is show gay movies all the time," to which Roger Ebert shot back, "Gene, that's ridiculous. Out of the 100 features this year, maybe 10 had gay-sensitive subjects." It took a few years for Gene to come around and discover and start reviewing LGBTQ films favorably. Of course, they had become much more mainstream by then.

Scud is beautiful at sensual exploration and the meaning of life, so there are a lot of naked guys all over the place. And I have to say, I approve.

5 – LA DOLCE VITA

I lost my innocence, if not my virginity, in Italy.

It was 1968 when I received a telephone call from an Italian journalist in Italy named Gian Luigi Rondi. I didn't know him from Adam. He had been reading about our Chicago International Film Festival as the place to find the new, young independent films and filmmakers. He had seen the racy Skrebneski posters and ads we'd been running in *Variety*, the ones that either shocked or outraged you.

Rondi asked if I would come to Italy and bring films to the famous Spoleto Festival. Up until this time, Spoleto had only been about music and theatre, and it was run by the famous composer Gian Carlo Menotti. So, this was an especially wonderful offer.

I brought the films and the filmmakers to Spoleto, and for me a whole new world opened up. I was introduced to Luchino Visconti, who was the president of this new cinema section, and Rondi was the artistic director. I was a kid and didn't even realize I was about to continue to meet film royalty from that day forward.

Spoleto Cinema led me to lend a hand to another Rondi festival, this one in Taormina, Italy, where he put me on a film jury with the heads of the Cannes and Tokyo festivals and film directors from three other countries. Then came Rome.

I was about to commence scouting films in Italy for my own festival in Chicago, and Rondi had quickly become my mentor before I even knew what a mentor was.

"You need a publicist to help you find the Italian films and meet the people," Rondi explained. This crazy/flashy PR guy was himself an outrageous queen, but everyone loved him for it. I know I did.

Guidorini Guido ran the Dino De Laurentiis Entertainment Group studios outside of Rome. They were booming at the time, filming the campy 1968 flick *Barbarella* starring the very young and sexy Jane Fonda, and with two Fellini films in production. It was an amazing adventure to witness what was going on at the new state-of-the-art facility. Dino himself was wild, old, and eccentric, with a fabulous actress wife, Silvana Mangano.

When I went on set to meet with the great Italian director Franco Zeffirelli in '67, he was directing *Romeo and Juliet*. He came over at one point and blurted, "Michael, you look so unhappy. What's wrong? I hate your dark hair. Come to the studio tomorrow and my girl Sheila Pickles will take care of everything."

So, I went to the studio as ordered. Sheila took me to the hair and makeup department, where they were painting the hair of all the cast members.

"Make Michael a blonde," Sheila said.

From that day forward and for 35 years, I was a fabulous blonde. The only one who couldn't seem to understand it was my father. When I came back from Rome that year, Dad met me at the airport, and the only thing he said was, "I thought Jean Harlow was dead." He never expressed a word about my being a blonde again. My mother remarked, "Well you were a blonde when you were born," so that subject was dropped for 35 years.

Zeffirelli, it turned out, was flamboyant and gay in every way yet at the same time a very Roman Catholic – and a genius. He brought youth to Hollywood movies. Remember, he created what Tom Cruise became in *Endless Love*. He loved using very young actors. His Romeo, Leonard Whiting, was only 17 at the time the film was shot, and Juliet, Olivia Hussey, was a mere 16 years old. Both of them wound up naked onscreen.

Franco's work was always described as opulent, extravagant, and baroque. He was, after all, taught by the great Luchino Visconti throughout his early years in film.

My screenings of new films in Rome started for me immediately. One day, Guido asked, "What are you doing tonight? Come to a party." That was where the fun began. It was set at Zeffirelli's villa outside Rome. I guess I really hadn't ever considered what a handsome guy the man was.

At one point, Guido said to me, "I need six young men for the big musical number" that he was planning for the evening's party. I had no idea what he was even talking about. My PR maven instructed me to take the car and driver and head to the central train station in Rome to rent six hot young boys for the party.

Yes, I guess that's just how they did it in Italy. This was fun.

Back at the party with the team of rent-a-boys, Franco instructed them to work with the party's musical and costume director, Damiano Damiani, a longtime Italian cinema writer-director. As to whom the party guests might be, I had no idea, but you always have a count and countess of some sort at all parties in Rome, as well as plenty of actors and actresses.

The main room of the villa looked out onto a large garden. The huge French doors all opened to the garden as well. It all became the stage for the show.

Imagine, if you will, that the audience is seated in the living room witnessing a live performance staged by Fellini's people along with Zeffirelli! I was already starting to be impressed. But I had no idea what was yet to come.

The show began with a red Fiat convertible driving past the window and stopping, driven by one of the railway station boys dressed as a sailor. The other five boys were positioned around the car also impersonating sailors. They all helped the star of the show out of the Fiat's front passenger seat. It was a giant woman singing some prerecorded Italian showtune, basically a fat man in a giant body suit looking like a bloated Anita Ekberg. She proceeded to sing and strip till she was practically naked, clad only in a G-string and pasties.

Remember that this person was still squeezed inside a huge fat suit, singing and dancing up a storm while surrounded by six faux sailors. It had evidently all been thrown together in a few hours. Very impressive.

Did I mention the star of the show was actually Damiano Damiani? Of course it was.

So how did it end? The six sailors stayed the night with the host, naturally. He had, after all, paid for them.

Franco remained outrageously gay until his famous car accident in 1969. Gina Lollobrigida, actress and photojournalist, was driving her Rolls Royce. Gian Luigi Rondi was in the back seat, Franco in the front. There was a young, unnamed man in the car as well. Gina smashed up the car. Franco, not wearing a seatbelt, flew through the windshield. He was badly injured and nearly didn't survive the accident.

From that moment on, Zeffirelli changed. He became more Catholic, including opposing gay and abortion rights. He still made relatively sexy films, but his fabulous good looks had to be reconstructed, and it seemed to have a profound effect on him. He was never the same, though he lived to be 96.

Meanwhile, I went on to become the official advisor for the Venice International Film Festival and for ten years was the U.S. film critic for *Il Tempo* newspaper in Rome. It all happened due to the influence of Rondi, a man more powerful than the Pope in Italy. Or so it seemed.

6 – I CALL FIRST

From the very start, my plan with the Chicago International Film Festival was to honor new filmmakers with their first film, a feature, short, student film, or documentary. Whatever. Trust me, this was a unique idea more than half a century ago.

At our first festival, we had two names who went on to amazing things, though they meant little at the time we featured them. One was William Friedkin, who would go on to direct *The Exorcist* in 1973. The other was Polish filmmaker Krzysztof Zanussi (Poland). Both directed documentaries. For Friedkin, it was *The People vs. Paul Crump*; for Zanussi, it was a short that was entitled *The Death of a Provincial.*

Fast forward two years. It's now 1967, and our festival scheduled *I Call First*, later renamed *Who's That Knocking at My Door*, by a young filmmaker named Martin Scorsese. It singlehandedly got us noticed.

Here's a poignant recollection from Roger Ebert: "In autumn 1967, I had been a film critic for seven months. I walked into 'the submarine,' the long, narrow, dark screening room knocked together out of plywood by the Chicago International Film Festival. I was 25. The festival's founder, Michael Kutza, was under 30. Everything was still at the beginning. I saw a movie named *I Call First*, later retitled *Who's That Knocking at My Door*. If I was sure of anything, it was that it was the work of a natural director, the American Fellini."

Scorsese would later note, "I sent my little film to Chicago. I couldn't go with it because I didn't have any money. It started as a short film I made as a student at NYU, then lengthened into a feature film on a shoestring."

The rest is history. It started a long Scorsese relationship with Ebert and the Chicago Festival, I'm proud to say.

In 1980, a film journalist, publicist, and aspiring screenwriter-director named Bertrand Tavernier was on our feature film jury. He was a delight, and always said, "One day I will direct a film and I promise I will give you the world premiere." Well, he did make his first film, and he lived up to that promise with *The Clockmaker of St. Paul*. I was very pleased, and we stayed friends.

Tavernier's films became increasingly famous, as did he. Eventually, when I wanted the latest one to premiere at our festival, Bertrand said, "Of course, but I'll need two first class tickets on Air France."

"We can't afford that," I assured him.

"Well, I guess I will head to the New York Film Festival instead," he replied.

We laughed. But he did. So much for loyalty.

Director Jan Troell of Sweden became a regular after his first film, as did many others like Mike Leigh from the UK with *Bleak Moments*. Michael Apted, Wim Wenders, and Patrice Chereau all became regulars, remaining loyal to the Chicago Festival with each premiere of their films. American director Taylor Hackford (Helen Mirren's husband) and I have remained close to this day. His feature *The Idolmaker* in 1980 was a stunning first film as was his *White Nights* in '85.

When I think of just a few of the directors who had their first showings in the United States at our festival, it's an impressive international list. It includes Oliver Stone, Joe Swanberg, and Faye Dunaway (USA); Maria Luisa Bemberg (Argentina), Alan Parker and Steve McQueen (UK); Armando Robles Godoy (Peru); Krzysztof Kieslowski (Poland); Walter Salas (Brazil); Amos Guttman (Israel); Vera Chytilova (Czechoslovakia); Jaime Humberto Hermosillo (Mexico); Leos Carax (France); Imre Gyongyossy (Hungary); Liv Ullmann (Norway); Dusan Makavejev (Yugoslavia}; and Scud (Hong Kong).

Other greats join the list each year.

How did I find these filmmakers? From the start of my festival, I traveled the world visiting film institutes and meeting with film historians and film critics, each time asking, "What's new in your country?" I would screen an endless number of films each year, and you know immediately when something feels unique to you and jibes with your sense of taste or style.

Then again, many of the films only submitted because their filmmakers had read about the Chicago Festival and wanted to be a part of the action. It was all part of the competition to be discovered – pretty much that simple. I always trusted my instincts about a film back then, and still do today. I mean, who else's instincts was I going to trust?

I can never forget George Christy, the entertainment columnist for *The Hollywood Reporter*, who had a regular column called THE GREAT LIFE. We would bring him in to Chicago every year to cover the opening night or a special gala. He always gave us great coverage, but there was a catch: he demanded not just a first-class airplane ticket on American Airlines, row one and seat A on the plane; he also insisted on a 24-hour car and a suite at the Four Seasons Hotel in Chicago. It was insanely expensive. Thank God he stayed for only two days.

Christy pulled the same gambit every year with the Toronto Film Festival, though probably with Air Canada and a different luxe hotel. It was there that he also hosted a lavish star-studded annual luncheon. I was amused when I attended and noticed that George had dedicated the main course to Canadian film and theatrical producer and entrepreneur Garth Drabinsky, who was later convicted and sentenced to prison for fraud and forgery. Good taste, George!

I remember an incident in L.A. one year with Christy. We were at the trendy Ivy Restaurant on Robertson Boulevard. An old lawyer friend, Jeffrey Jacobs, approached.

"George," Jeffrey said, "I want you to come over to my table and meet my associate. She's going to become the biggest thing on television. Her name is Oprah Winfrey."

"I'm not going to anyone's table, you bring her to me," George replied snootily. Jeffrey was the president of Harpo Productions, which was Oprah's company. George got his way, and Oprah came to his table as a humble young lady to kiss his ring. Probably never happened again.

I also remember that each year, George demanded that our photographer's photos of the events be brought to him personally as soon as they were shot, as he was on a tight deadline and would need them the next day. One year, the guy shooting them, a nice guy, young, cute, said to me, "I don't want to take the photos to Mr. Christy anymore. Can you have someone else do it?"

George insisted that the photos be brought to him at his hotel. The young photographer told me later, "I don't like going to his room. He makes me sit next to him on the sofa as we go through the photos. He's only wearing his bathrobe, and I'm very uncomfortable when he opens the bathrobe as I'm showing him the photographs. I don't want to go there again. Please have someone else go."

We did have someone else do the job. That same young man took his own life a few years later. It was very sad.

Speaking of Jeffrey Jacobs, one year I was sitting at the Palm Restaurant in Chicago with real estate tycoon Jerry Wexler. I was really hot at that time, with a successful festival and many Hollywood celebrities attending. My face was even painted on the wall in the restaurant along with all the fancy names in the city.

Jeffrey approached our table with his legal partner, Saul Foos, saying, "I have a project for you. I think you might like it. It would be important, and I think it will change your life. I want you to do movie reviews three times a week for ABC News in Chicago." (Mind you, this was before the days of *Siskel & Ebert*, way back at the beginning before critics on TV became a thing.)

"Well," I told Jeffrey, "I'll give it some thought."

That wasn't good enough for Jeffrey.

"No don't," he urged, "just say yes, and I'll have a speech instructor and a TelePrompTer coach come over to get you rolling, so you can do this. You would be inserted during the news every other night."

I participated in the lessons with the instructor and did a few auditions. But I didn't feel like I had the time to give this what it needed to succeed.

"Jeffrey, I have the festival to run," I told him, "and I must be traveling to Berlin and Cannes and Venice seeking the films. How can I do these movie reviews every other night here in Chicago?"

"Well," he assured me, "you can't go to those places anymore."

The thought of being tied down didn't work for me. The other part I was uncomfortable with was the fact it was tightly scripted. I would write my review and basically read it from the PrompTer. I'm at my best off the cuff and extemporaneously talking about films, not reading a review. I walked away from what was clearly a great opportunity. But the way I saw it, I really had no choice.

Another similar missed opportunity came up when gossip columnist Aaron Gold of the *Chicago Tribune* died of AIDS in 1983. I was asked if I would like to do a daily gossip column, as I "knew all the stars and behind-the-scenes stuff."

"If I did what you want," I countered, "you would need a full-time lawyer on staff just to cover me and my mouth. It's simply too dangerous for me and for you. Sorry."

In 1975, I was planning a retrospective of Italian film director Pier Paolo Pasolini. We would be showing six of his films and he'd be there in person for the tribute. Everything was in place, until I received a phone call at five in the morning from his agent, informing me that Pasolini had just been murdered by a young man he'd picked up at the train station in Rome. Pasolini's hitting on him didn't go well.

I quickly rang up my festival chairman Charles Benton to tell him that Pasolini had been killed.

"Does this mean he's not coming?" he asked.

"Yes, Charles, that's what it means.

"Well," he replied, "that saves us an airplane ticket."

The film retrospective nonetheless went forward without the director. Charles was actually my first chairman of the Chicago Festival. I've had an interesting array of chairs over my fifty years. Men, women, gay, straight, black, white. From egomaniacs to warmhearted fatherly leaders, they all had a single goal: to keep the dream alive.

Charles was my new father. He'd called me a few weeks after my father died, having been following the festival and feeling that I needed help. He offered to put together a board of city leaders in all fields in Chicago to move the organization forward while letting me do what I did best: find the films and the celebrities.

He made that board happen and also put in place a wonderful woman whom he knew from his political dealings. Her name was Judy Gaynor, and she quickly became our festival's mother. Every festival needs one.

7 – IN OLD CHICAGO

For some reason, I wasn't nervous at all meeting one at the most famous politicians in the United States: Richard J. Daley, strongman mayor of the city of Chicago and known as the last of the big city bosses. He's the guy that's famous for saying "Shoot to kill!" during the 1968 Democratic National Convention.

Mayor Daley agreed to see me and hear who I was and what I wanted for his city and my dream festival. My father, who had just died, was the major money behind the festival. I needed help, so I went to my new father.

The mayor didn't seem to know that I was partially responsible for the move to disband his film censor board. Remember that Chicago at the time was a very Irish and Catholic city, so Daley had put in place a board of nine widows of Chicago policemen to review all films and rate them as to whether they were acceptable for public viewing. This is before Jack Valenti and his Motion Picture Association of America (MPAA), which rated movies on the G, PG, M, R, and X scale.

So anyway, here we had these nine old ladies with no film experience whatsoever judging the merits of films. I had to personally drag in every one of our features to be screened by these women. I recall there was a $100 screening fee as well.

The city of Chicago had set up a screening room in an actual courtroom. Since my films where basically foreign, that already meant they were automatically somehow "pornographic" in their view. I wish I were joking.

After they screened about ten of my festival films, they became completely bored with the process and the subtitles and the non-sexy content. We decided to make them all "adults only" just to stop this nonsense. I found the whole thing ludicrous and mentioned it to a few newspapers. It became embarrassing to the city, and slowly the censor board was disbanded. I hope I had something to do with that.

Of course, Mayor Daley had problems of his own with a film called *Medium Cool* by director Haskell Wexler, a behind-the-scenes look at the messy Democratic Convention and the police beating the young people pro-testing the mayor and the Vietnam War. From that day on and that film on, no films could ever be shot in Chicago again without script approval from City Hall.

With all that background information, here I was asking the boss, as he was known, for a favor. I needed help from the city to survive. Mayor Daley listened and finally said, "Ya know, those kinds of films you show can lose me votes."

"Yes, I understand," I replied, "but we are putting a very positive spin on the city. I could use your help with theaters, hotels, and hospitality."

The mayor thought for a while and said, "Give the kid McCormick Place."

Now, this was a giant convention center nowhere near where our audi-ence hung out. It was also far away from anything else of note in the city, I pointed out. But instead of responding to me directly, he again barked out the order, "Give the kid McCormick Place." (This was where the convention center was.) Daley also introduced me to Frank Sullivan, his press secretary and close assistant in most everything.

"Frank will take care of you," he said. "I'll help you, but I don't want my name on anything, remember that."

Daley was afraid of me, but I amused him. His son, also named Richard and who would later also become mayor of Chicago, hated me. You win over some people, others not so much.

Sullivan was great at twisting arms to get things done. And we did actu-ally use McCormick Place after all. Unfortunately, nobody came because the venue was so far away, but at least we pulled it off and survived another year. I went to City Hall after that to promote the festival and the city with new events.

Since Chicago was the home of the first roller skate manufacturing com-pany (established in 1905), why not sponsor a giant roller skating event? The idea was to close down Lake Shore Drive on a Sunday and have peo-ple skate from one end to the other, naturally while wearing Chicago Inter-national Film Festival t-shirts.

This was unfortunately instantly shot down by a new player I was about to meet, Colonel Jack Reilly, the event planner for the city of Chicago. (I had no idea where the "Colonel" part came from.) It seemed that I was about to step on his toes, and he didn't appreciate it.

Col. Jack's most famed idea (one that remains today) was to dye the Chicago River green on St. Patrick's Day. He didn't go for our roller skating idea, nor another one that called for organizing a giant kite festival over Grant Park with all the kites sporting our festival logo. Both were immediately shot down by The Colonel. The problem with the kites was the fact there was an airport nearby, and dodging kites could cause planes to crash. Bad idea.

Parades were more the speed of both The Colonel and Mayor Daley. The public evidently loved them, too. Meanwhile, Col. Jack found me and my crazy ideas to be a threat.

One thing that I did promise Old Man Daley in return for his assistance behind the scenes was to always promote film production in Chicago. We would build it into all national and international ads for the festival, working closely with Frank Sullivan to develop a Chicago Film Office of sorts. It was all about playing well together in the sandbox. I've always been good with sand.

I was very good at using smoke and mirrors. I used that expression a lot because throughout the history of the Chicago Festival, we never had any money. It was always a matter of struggling to make it appear we were something gigantic, this massive and lavish thing.

In reality, nothing could have been further from the truth. We were always in debt and on the verge of dying. It was a mess. There were lots of times I didn't think we'd pull through. But we always survived, somehow. So did I. I would do graphic artist stuff on the side and scrape together a salary from the festival. But the truth is that every cent I ever had, I poured back into it. That was my foolishness. I believed in it that much.

And you know what? If I had it to do over again, I wouldn't change a thing.

8 – STARDUST MEMORIES

Did you know that in 1977, Lina Wertmuller became the first female director ever to be nominated for an Academy Award®? She's a fascinating, trend-setting Italian director who directed some 30 films while also having a hand in innumerable television productions and live theatre. She worked up until the day she died at 93, in December 2021.

One day in 1990, I was having lunch with Lina in Rome on her terrace overlooking the Piazza del Popolo with Guglielmo Biraghi, the film critic, playwright, and director of the Taormina Film Festival. He was trying to convince Lina to be on a film jury at the Sicily Festival. She agreed to do it on the condition that she have a chance to see all the films in two days one after the other rather than spending days watching films and attending parties. She was concerned about getting back to work in Rome.

Lina then mentioned she was finishing her new feature film *Saturday, Sunday and Monday*. I asked if she would be interested in premiering the movie as our opening night entry in Chicago. It starred the legendary Sophia Loren and was excited at the prospect of having it launch at our festival.

"Well," Lina said, "you would have to convince Sophia's husband Carlo Ponti, the producer of the film."

"How do I do that?" I asked.

"You would go visit the two of them. They are in California at this moment at their ranch."

It was to be the start of a very long friendship with the Loren-Ponti's. I went to Los Angeles and drove a few hours out to Hidden Valley. Don't ask me where that is. As the name implies, it's hidden. I eventually did find the place, driving up the long and winding driveway through many beautiful Italian artifacts in the gardens. Finally, you make it up to this Italian villa.

I was welcomed by Mr. Ponti himself, the man who discovered and married Sophia and of course guided her career all these years. I told him all about the Chicago Film Festival and how we could honor Sophia and roll out the world premiere of his and Lina's latest film. We had coffee and he began considering the idea before asking, "Would you like to meet Sophia?"

At that moment, a vision came walking in the distance through a doorway toward me and all I could see were legs, amazing legs. She sat down in front of me and we started to talk about Chicago and the festival, and I was just in jaw-dropping awe at this incredible beauty seated three feet from me. Sophia was tan, tailored, and altogether exquisite.

We started working out the details of her coming to Chicago and receiving a lifetime achievement award or whatever we were going to call it.

"Well, you know, I have to bring Carlo and my children with me, and Lina must come from Rome with her assistant," Sophia noted.

"No problem," I assured her.

We started working out the details. It took several months, but the date was chosen along with the location for the gala after the film along with some fun, unique ideas to make it a memorable evening for Sophia, Carlo, Lina, and the kids.

I started a dialogue with Sophia. She wanted it all to be perfect in every way. She had not been honored in the United States like this and was excited. What would be the right hotel for them? What would she wear? I offered our official hotel, the Park Hyatt, which was brand new, or the Four Seasons, popular at that moment, or possibly The Drake, which was steeped in Old World charm.

There was no Peninsula Hotel at that time. Interestingly, The Drake was the winner. Sophia wanted that yesteryear feeling and lots of antique furnishings and a view of the lake. Sophia also liked the lower-level elevator access to avoid the public while entering and exiting the hotel.

A few weeks later, Sophia called and asked, "How many stairs must I walk up or down at the gala in my dress?" Then we talked about the gala food menu. The lady was a stickler for the details.

I had a fun idea, which Sophia loved. She always claimed that her success was due to spaghetti. Seriously. "Everything you see I owe to spaghetti and more," she claimed.

"Well then," I offered, "let's serve your recipe of spaghetti sauce on one of the dishes at the dinner." Sophia loved it, gave me the recipe, and I had Joey Mondelli of Kelly Mondelli's restaurant make it. But first, she had to approve it. Joey made it and I flew out with a gallon packed in dry ice back to Hidden Valley. True story. All turned out fine.

Then there was the dress. Sophia had Givenchy design it. At our opening night event, she asked, "What do you think of It?" I mean, of course it was a work of art and showed enough legs in front to make it up and down the stairs comfortably. It came in at a mere $25,000. And in case you were wondering if Sophia did her own makeup and hair, the answer is yes.

Meanwhile, at the gala dinner that night, Sophia was simply a dream, meeting guests with charm and very cordial to everyone. Just a class act all around. However, when she was seated and eating dinner, if anyone dared to come near or interrupt her personal time, they would receive a look from her that told you steer clear of me or die. Only an Italian woman can deliver that look so powerfully. Trust me, I know, having been raised by an Italian mother (and a doctor no less).

--

Sharon Stone was a model and TV actor when Woody Allen cast her in her first film, *Stardust Memories*, in 1980. (She played the role of "Pretty Girl on Train.") But it was Paul Verhoeven's film *Basic Instinct* in 1992 that of course put her on the map in a significant way. You no doubt remember that scene when she uncrosses her legs, revealing much more than is typically seen in your usual Hollywood detective flick.

In Chicago, our festival posters by Victor Skrebneski were really in demand but needed a different twist from naked men and woman and wet, torn t-shirts. There are only so many of those you can do before it gets a tad repetitive.

Victor wanted a star for the next poster. *Playboy's* Marilyn Grabowski was very close with Sharon, who knew Skrebneski's work and was willing to come do the festival a favor as long as she could meet Victor. This was in '92, after she'd become an instant star for *Basic Instinct*. Her demands were simple: a private plane, seven bodyguards, and the whole trip must be kept secret to avoid any press.

Sharon knew she was a hot property because of the film. That's why she wanted to keep her arrival as well as any moves around town secret. As a result, she refused to fly commercial to Chicago, which was a problem for us that she solved. She found a private jet that would bring her to the city for the poster shoot.

I got the limo service to drive up to the plane as it was landing at the private area of the airport. As she was deplaning, Sharon saw maybe ten guys with cameras and binoculars eyeing her.

"How did these people know about my secret arrival?" Sharon asked accusingly. The limo driver admitted that he'd told a few friends.

"Fire him!" Sharon demanded, right on the spot. I did.

Another limo and driver were immediately summoned. I left my car at the airport and stayed with Sharon for the drive to the hotel. She was in great spirits from that moment on.

We did the photo shoot with Sharon. Victor's people had hair and makeup people, but Victor dismissed them and decided he wanted to create a new look for her, so he did it all himself. A new Sharon Stone appearance happened on the spot, a look she kept for several years.

After the shoot, Sharon did a cooking demonstration over at Planet Hollywood. Why? Well, that's how she got the private jet. Clever woman.

Her stay continued in Chicago with a new young man she was involved with at the time. The poster was amazing, and Sharon's participation led to other stars willing to do posters for us. These included Sigourney Weaver, Faye Dunaway, Shirley MacLaine, even Steven Spielberg.

--

Lauren Bacall was plenty difficult from the moment she arrived in Chicago to receive our Lifetime Achievement Award that chilly October of 1999. We set her up in a fabulous penthouse at the Hilton Hotel and Tower, but she hated it. I went to calm her down. It seems her problem was that the staff didn't know who she was. This really irritated her as she couldn't push her weight around.

Yes, Bacall is an older star and as such expected star treatment. But she wasn't going to get it from maids and housekeepers who barely spoke English and hadn't a clue that she was anyone.

Our beloved Robert Allegrini once again came to the rescue. He moved Bacall to a different suite, and a different, English-speaking staff was filled in about who she was. He sent flowers and a nice bottle of champagne, all designed to calm her frayed star nerves.

The evening tribute itself, meanwhile, started out smoothly. Lauren looked terrific. We had some surprise guests and film lovers to speak on stage, including a film critic who recited the famous line from *Casablanca*, "Oh, maybe just whistle. You know how to whistle, don't you Steve? You just put your lips together and blow." He then got the entire audience to whistle. It was fun, and the honoree enjoyed it.

The emcee, Bill Kurtis, was excellent as always – as long as he stayed on-script. Unfortunately, he didn't. He foolishly decided to tell the audience, "Today is Miss Bacall's 75th birthday. Let's all wish her a happy birthday!"

From the first row of the theater came Bacall's yell, "YOU ARE DEAD!"

From that point on, Lauren Bacall was a hostile guest. To make the evening perfect, I even brought in her favorite writers Betty Comden and Adolph Green, who did the Broadway musical *Applause*, to praise her on stage and calm her down. But nothing helped because she was just plain pissed off. When she finally came on stage to speak, she kept referencing the age and birthday announcement.

"I'm opening on Broadway in a few weeks," she said, "that is, if I'm not too old." It went on like that. She did have some great memories of being on the stage here in Chicago years before and said she'd love to come back and work at the Steppenwolf Theatre, if only they would ask.

At the gala dinner that followed, Bacall was of course at the main table with the festival board members who paid to sit near her. They unfortunately kept clawing at her and boring her with Bogie & Bacall questions. The orchestra was also too loud for her, and it got to the point that the music and the people became too much, and she wanted out.

"She said, 'Get me out of here, NOW!'," I told a few of our board members.

I eased her out quietly from the main ballroom toward the limo when more fans came running at her with cameras and autograph books. She went ballistic.

"Get these gay hairdressers away from me!" Bacall demanded. "Who do they think I am, Angela Lansbury?" She should have been so lucky.

We finally got her in the car and back to the hotel. But she wanted completely out of Chicago, immediately if not sooner. Let's just say it wasn't a good time for Miss Bacall, since she suddenly felt old that evening. (Thanks Bill.) it was her birthday, and she wasn't happy about the adulation and being alone in a city she no longer felt good in.

--

Faye Dunaway came to our festival on two different occasions. The first was for her 1988 feature *Burning Secret*, which we presented as an opening night premiere that year. She came back in 2001 with a short film she directed entitled *The Yellow Bird,* and we turned that evening into a full-blown gala Career Achievement event.

Faye doesn't like to do interviews or even giant events honoring her. She insists on keeping her personal life, personal. I guess that I somehow convinced her to let us do a stunning evening in her honor complete with surprise guests.

But let's back up to the first time she arrived at the Chicago Festival thirteen years before.

We had been prepared for a very difficult star after hearing all sorts of horror stories about her demands and just being a Hollywood diva bitch. I remember that Faye had to have a white limo pick her up the airport. Not easy to find, except for weddings or African American funerals.

No sooner did she arrive at the hotel than she immediately launched into a fit because she couldn't get the TV to work! She screamed at the hotel management, "If this doesn't work, what else is not going to work in this place?" About this time, the hotel received a call from the limo company saying that Miss Dunaway had taken the remote control from the limo and they would like it back. Whoops.

All was calm after that.

I found during my half century with the Chicago International Film Festival that real stars like Faye – a brilliant actor in some important films like *Bonnie and Clyde*, *Three Days of The Condor, Chinatown, Network,* and of course *Mommie Dearest* – tend to be perfectionists. That's what she is, all right, along with insecure and (at this point in her career) quite alone. She wanted what she wanted, when she wanted it.

If you treat a Faye Dunaway like a star, fussing over her and indulging her every whim, you'll generally do just fine. But if you fuck something up with her, you're basically dead.

Our tribute evening that year with her was magical, and she actually seemed to enjoy it. Roger Ebert, the esteemed television and newspaper film critic, was the guest speaker, and he pretty much saved Faye's career with his rave review of *Bonnie and Clyde*. The critics and the public had panned it, but Roger's praise brought it back onto the market and into theaters after it had initially bombed. The film was reborn. Faye knew that and was thrilled to thank Roger in public from her heart.

While Faye was thanking me for the honor of her Lifetime Achievement, however, she continually forgot how to pronounce my last name "Kutza" and kept saying "Chutzpah" instead. It was funny, because, I mean, it does sum me up. I took it as a compliment. But it was still a bit annoying.

I'd seen Faye in Cannes and at festivals relaxed and enjoying herself. But I also witnessed the opposite at Cannes, with Ferocious Faye on frightening display.

Faye was noted on the official Cannes poster that year in a glamor fashion shot of her by a onetime lover of hers, director and photographer Jerry Schatzberg. In tribute to Faye, they had restored Jerry's 1970 film *Puzzle of a Downfall Child* that starred Miss Dunaway.

The premiere screening was in the morning in front of about 1,000 press members and guests. But it turned out the festival couldn't start the film since they didn't have the correct key for the digital projection system.

Uh oh. Trouble.

This was such an important moment for Faye and Jerry, and it was held up for over an hour. It of course brought out the horrific side of Faye, and frankly I couldn't really blame her. Man, it's tough being a movie star.

Oscar® winner Robin Williams accepts an award at
our Chicago International Film Festival in 2004. (Robert Dowey)

Chicago TV entertainment reporter Bill Zwecker
interviews Robin Williams. (Robert Dowey)

(left to right) Israeli actress Gila Almagor, Greek actress Irene Pappas, and French actress Isabelle Huppert, showing off the international flavor of our Chicago Festival. (Robert Dowey)

Entertainment lawyer and Chicago Festival board member Randy Crumpton. (Robert Dowey)

The actress Pam Grier at our festival in 1998. (Robert Dowey)

Film critic and author John Russell Taylor (left) with
Academy Award® winner Liza Minnelli. (Robert Dowey)

Actor Lawrence Fishburne of *The Matrix* fame. (Robert Dowey)

Me and the great actress Kathleen Turner snuggle up. (Robert Dowey)

Actors Kelsey Grammar (left) and John Mahoney of
Frasier fame. (Robert Dowey)

At an Oscar® party, festival staffer Kelly Moseley met then-Senator Barack Obama.
Rumor has it that the senator continued to progress in politics. (Robert Dowey)

Director Richard Attenborough was a good friend
to the Chicago Festival. (Robert Dowey)

Deborah Harris poses in front of a giant blow-up of the provocative cover of
my book "Moving Pictures!", dedicated to photographs shot by unrivaled
photographer Victor Skrebneski. It features some spectacular portraits
and posters from the Chicago Festival. (Robert Dowey)

Multiple Oscar®-winning director Oliver Stone (left)
and Jim Belushi. (Robert Dowey)

Me (left) with the Austrian-born actor Maximilian Schell. (Robert Dowey)

(left to right) The wondrous Italian director and screenwriter Lina Wertmuller with me, the beautiful Italian Oscar®-winning actress Sophia Loren, and Sophia's husband, producer Carlo Ponti. (Matt Gilson)

Lauren Bacall is not pleased by what she's hearing at our Chicago Film Festival in 1999. (Robert Dowey)

Judy Gaynor (left), then executive director of our festival,
with two-time Oscar® winner Jodie Foster in 1996. (Robert Dowey)

Faye Dunaway (left), an Oscar® winner for *Network* in 1977,
with Roger Ebert at our festival. (Robert Dowey)

Ellis Goodman, onetime Chicago Festival board chairman,
is surrounded by the Hawaiian Suntan Girls at the
Cannes Film Festival. (Cannes Festival)

French director Francois Truffaut, a three-time Oscar® nominee,
with French actress Fanny Ardant at our festival. (Steve Arazmus)

Two-time Oscar® winner Shelley Winters (left) and book publicist
extraordinaire and friend Lynda O'Connor share a bite in
the Pump Room. (Chuck Berman for the Chicago Tribune)

Robert Englund (best known as Freddy Krueger in the *Nightmare on Elm Street* films) cuts an amusing profile at our festival. (Steve Arazmus)

I stand idly by as the distinguished film writer and critic
Charles Champlin (far right) does a quick fix on Walter Matthau's
tux (center) at our tribute to Jack Lemmon. (Steve Arazmus)

Actor Richard Gere is all smiles at our Chicago Festival. (Robert Dowey)

Our beloved festival executive director Suzanne McCormick and
I share what looks like a special moment. Actually, all my moments
with Suzanne were special. (Robert Dowey)

9 – A FEW GOOD MEN

Honoring Steven Spielberg wasn't easy. But it sure paid off.

I was raised on Hollywood movies, and Spielberg was a hero of mine. I remember in Cannes back in 1974 when he premiered his first feature film, *The Sugarland Express*. We had screened his made-for-TV movie *Duel* in Chicago three years before. This was different. This was big time.

So, there we were in 2006, and I was determined to honor him. I called my mentor Paula Wagner, the fabulous agent, producer, and wonder woman in my life when it comes to Hollywood. She said, "I'll call DreamWorks, but you'll have to convince Marvin Levy why this tribute should happen."

Marvin Levy had been Steven's publicist, sounding board, and best advisor going back forever. He's listed as just a publicist, but his 40 years-plus relationship with Spielberg proved he knew what worked and what didn't for Steven.

I headed to Burbank and met with Marvin at Amblin Entertainment/DreamWorks Studios on the Universal lot. He's a wonderful film maven who was expert at protecting Steven's interests. He liked the idea of honoring him at our Chicago Festival if it could fit into his busy schedule of three film productions and his children's school holiday. I promised we would schedule the event around anything he wished and would feature special guests and surprises.

"No surprises!" Marvin ordered. "I want all the details and planned guests. We'll prepare a film clip reel of our choosing."

I think Marvin knew I was a fan, but not a star fucker. We started putting it all in place for his approval. We had the lead actor from *Jaws*, Roy Scheider. We got the world's leading expert in discovering dinosaur bones from the University of Chicago, paleontology professor Paul Sereno, to cover the authenticity and success of *Jurassic Park*. We even enlisted an Auschwitz survivor to tie in with *Schindler's List*.

We also gathered video clips from those who couldn't make it in for the tribute, including Tom Cruise, who had just completed two films with Steven: *War of the Worlds* and *Minority Report*. Tom shot a video from his home in Telluride, Colorado wishing Steven the best. It was great, and Marvin loved it all. But the audience was convinced Tom couldn't be there because, I mean, he'd sent the video from Colorado.

The big day came. Steven arrived with a slight problem, having suffered a wasp sting on his forehead and wanting it covered before the evening tribute. We called the makeup expert in town, Marilyn Miglin, and she sent over two ladies at a cost of $700 to fix him up. I had Skrebneski at the hotel to shoot Steven for an official poster as well.

Everything was in place. It was time for the event. It was a sellout. Very exciting. The emcee as always was Bill Kurtis and he was in great form. Those who paid tribute to Steven on stage were all perfect and touching. Steven seemed thrilled.

Then came the moment I would present him with our Gold Hugo. I made the usual speech expressing my admiration to the honoree and reached to grasp the statue from beneath the podium, but it wasn't there. I stumbled in disbelief.

By this time, Steven was standing next to the stage when suddenly someone emerged from backstage clutching the statue. It was Tom Cruise, live and in-person. Steven was blown away. So was I.

I was so relieved that the charade of never telling Marvin Levy that I had kept this surprise from him didn't blow up in my face. Cruise, meanwhile, was terrific, asking Steven "How could I not be here for your big night?"

Spielberg would later write a note to me that read, "I was so moved by an event so full of surprises and deeply moving remembrances. How you ever managed to secretly pull Tom Cruise from his wife and family to present me the Gold Hugo was testimony to your uncanny ability to pull together such successful events.

"Thank you, Michael, for making me feel so relaxed and unprepared for one of the most emotional evenings of my life." It was a remarkable moment in Chicago International Film Festival history and mine, too.

I was hugely relieved. The thing was that Spielberg's people were very hands-on and needed to know everything down to the most minute detail. But Tom forbade anyone, staff or otherwise, from knowing the surprise was going down. Thank God it all went off so beautifully.

--

I grew up on actor Charlton Heston's epic Hollywood films like *The Ten Commandments, Ben-Hur,* and *The Greatest Show on Earth.* And then came a whole slew of trashy disaster films including *The Towering Inferno* and *Earthquake.* That's not to mention the *Planet of the Apes* movies and *Soylent Green.*

Heston was impressive and audiences loved him. So in 1976, I invited him to Chicago. He's actually from Evanston, Illinois, a few miles north of Chicago, and attended Northwestern University in Evanston. He liked the idea of a tribute and a retrospective of his films because it would also give him a chance to see his mother back in Evanston while in Chicago for the Festival.

He arrived in Chicago, and we took him to our official hotel that year, The Tremont. He had the penthouse suite, and all was well until he informed us that, at six-foot-three, he was too tall for the penthouse suite bed. It took some work at the hotel to find another bed for him before he came back for the evening.

We presented Mr. Heston on stage at the Uptown Theatre, a magnificent, giant movie palace with a capacity of 4,500. He was hosted on stage by film teacher and historian Arthur Knight. Arthur was a big deal back then with his monthly column in *Playboy* entitled "Sex in the Cinema" and his film classes at USC in California.

The festival had two dramatic squared Le Corbusier-style chairs placed on the stage, both great looking and very comfortable. However, the only problem was that they were made completely of foam rubber covered in fabric. So while they looked solid, they actually weren't.

Mr. Heston (I don't feel comfortable calling him Charlton) was warned off-stage, "Do *not* sit on the arms." So what happens? Moses comes on stage and immediately decides to sit on the arm of this seemingly solid chair. He falls flat on his ass, stunned for only a split second before instantly pulling himself up and comfortably sitting in the chair without a word. That's a pro and also a good sport. He filed no lawsuit, thank God.

On the screen at the Uptown, we presented clips of Heston's films, while the retrospective of his full-length movies was taking place at the Biograph Theater further south from the Uptown. The retrospective was a Charlton Heston Marathon for 24 hours. You could stay all night. Some students even brought sleeping bags. We served free coffee and sweet rolls for breakfast for those who made it all the way through.

We had the 35-millimeter prints scheduled so tightly that we knew exactly when key scenes would appear. Mr. Heston said he wanted to appear at The Biograph when the parting of the Red Sea took place in *Ten Commandments*. That was 3 a.m., and by God he came to the theater and greeted the audience in the middle of the night. Now that's an actor dedicated to his craft.

--

I received a call one morning in 1975 from Burton Kanter, the cofounder of the major Chicago law firm Neal, Gerber & Eisenberg. He offered me the world premiere of an important new film. The firm had just helped complete the movie using a new form of funding, a tax loophole of sorts. It was going to save Hollywood, and in this case Paramount Pictures, with new tax-saving shelter strategies.

The book *One Flew Over The Cuckoo's Nest* was also a play on Broadway starring Kirk Douglas in the Sixties, then was revived again in the Seventies by Kirk's son Michael Douglas and made into the Oscar®-winning 1975 film directed by Milos Forman. I was thrilled at the opportunity to roll out the world premiere and have the cast come to our festival in 1975.

The *Cuckoo's Nest* star Jack Nicholson was a joy to host that year. His needs were simple. He stayed at the Playboy Mansion on State Parkway in Chicago. *Playboy* magazine and its boss Hugh Hefner had an elegant dwelling with everything their heart may desire for celebrity guests like Jack. He also brought along his own marijuana stash. Meanwhile, we housed the rest of the cast at the Drake Hotel.

The evening's premiere began with an elaborate pre-party in the lobby of a sponsor's department store before we moved the guests to a massive movie palace up on the North Side, The Granada Theatre seated 3,400 people. It was French Baroque, with influences of Versailles and the Vatican. The giant stained-glass window was five stories high.

So now we added to this picture movie stars and 3,500 people, a hundred over the limit. Some audience members were literally seated on the floor in the aisle. Others were standing. We jammed every last body in there that we could.

You can pretty much guess what happened next.

Someone called the fire department, the cops, and the bomb squad to stop the show as a crowd hazard in the making. I started to panic and turned to Irv Kupcinet, the on-stage host of the evening and the guy I always had nearby to get us out of a fix. I enlisted Irv to talk to these guys who were demanding that we clear the house.

Kup, as he was largely known, explained to the officials that if we can-celled the screening there would be a riot. He felt it would be in everyone's interests and the safety of all concerned to let the show go on. It did. Kup had that sort of power. God bless him.

Next problem: the 35mm film had just arrived from the lab and was still wet, which caused it to go slightly in and out of focus when the heat of the projection bulb hit the film as it was being projected. This was driving the film's director Forman insane. He was screaming to focus the film from his seat in the auditorium.

Finally, Forman decided to leave his seat and sprint up the seven floors (yes seven) to the projection booth. The projectionists had locked the ele-vator, so Milos couldn't get up there except to climb the stairs. He finally reached the projection booth metal door and started pounding on it while screaming, "You must focus the film!"

"Get the fuck away from this door or we'll shoot you!" came the threat-ening reply.

Milos was unaware that the Mafia basically owned the projectionist's union, and you don't mess with the boys. He backed off. The filmed played on, and it seemed the wetness and blurring impacted only some of the reels.

The film completed, Nicholson walked on stage with Kup to a ten-minute standing ovation. What an evening. Thank God Jack was high all night and therefore unaware of anything negative that was going down. His costar Louise Fletcher and producer Saul Zaentz were ecstatic.

--

I had just seen the 1992 film version of the David Mamet play *Glengarry Glen Ross* starring Jack Lemmon, Al Pacino, and Kevin Spacey. I suggested to my chairman at the time, Yale Wexler, that we should make Lemmon the honoree that year. He'd had a terrific film career with *Some Like It Hot, The Apartment, Mr. Roberts, The China Syndrome,* and *Days of Wine and Roses.* He would be perfect for a Lifetime Achievement Award.

Yale was an interesting character. He was one of three brothers in Chicago. Jerry Wexler was a real estate tycoon who owned a good percentage of the fashionable Gold Coast area condos and hotels. His older brother was Haskell Wexler, a two-time Oscar®-winning cinematographer. He'd had a hand in making some terrific films, including *Who's Afraid of Virginia Woolf?, One Flew Over The Cuckoo's Nest, Coming Home,* and of course *Medium Cool* (which really stirred up old Mayor Daley).

The youngest of the three brothers, Yale wasn't exactly the family black sheep, but he was an actor, with some stage, TV soap operas, and a few films to his credit. He sometimes invested in plays. One was *Glengarry Glen Ross*, and his buddy was the producer. So, Yale made a call and started the ball rolling. We got Jack. Also coming onboard for the tribute were Walter Matthau, Jack's best friend and fellow actor who had made eleven films with him (memorably including *The Odd Couple*); Kevin Spacey, a great Lemmon admirer; and famed director Robert Altman. Altman had just used Lemmon in two films, *Short Cuts* and *The Player*.

Matthau was a riot talking about Jack's dark side, something he doesn't have. Spacey spoke about how he'd tried to meet Lemmon a lifetime ago in an encounter in his high school, pushing for an autograph. One of the great trivia moments from the *Glengarry Glen Ross* film was that these high-strung real estate salesmen in the film who are trying to stay alive in a cut-throat office utter the word "fuck" some 42 times, which was a record at the time for a feature.

I had never met Lemmon before this, but he was an easy, friendly, and sincere man. The tribute and the outrageous comments from his fellow actors brought him to tears of joy, as you can see by the cover photo. Of course, he's also an actor, but I believed the tears were real and not like John Travolta's at his tribute, which he could turn off and on seemingly at will.

We always did the award presentation for the TV cameras at noon, so they could get their B-rolls and all that crap for the news. The honoree would give the speech just for camera. In the evening, when the actual event took place, they would get the award properly.

So, when Travolta did his noon thing, he started tearing up about how important this whole thing was to him. I was very touched…until he teared up again at the same exact moment in the speech in the evening.

What can I tell ya? The man's an actor.

--

I grew up watching Robin Williams on television on *Mork & Mindy* and then later brilliantly on the big screen in more than fifty feature films including *Moscow on the Hudson*, *Good Will Hunting*, *Dead Poets Society*, and *Mrs. Doubtfire*.

But it was later in Williams' career when I was unnerved by his performance in 2002's *One Hour Photo,* which I saw at the Locarno Film Festival in Switzerland that summer.

I called the film's director, Mark Romanek, who ironically once volunteered at the Chicago Film Festival in our office when he was a young man. Mark had in 2010 directed another film that also still haunts me, *Never Let Me Go.* I asked him in 2004 to approach Robin about coming to Chicago to be honored. He said yes.

Summer gala in Chicago confirmed, Robin brought his entire family on a private jet, as they had planned to fly the moment after the gala to Paris to watch his buddy Lance Armstrong win the Tour de France bicycle race.

"As he's riding, the French people were blowing cigarette smoke at him and saying, 'Good luck, Cancer Boy!'," Robin said. Classy bunch, those French.

Sitting next to Robin at the tribute, I was able to see this rapid-fire humor in action. I introduced him to the various guests at the head table and he would chat and they'd casually tell him things happening around Chicago. It turns out Robin had a little-known connection to the city – and to me. He was born at St. Mary's Hospital, where at the time my surgeon father was the president. Crazy irony. He was raised in Lake Forest, probably the richest suburb in Illinois and about 40 minutes from Chicago.

What Robin did ever so subtly was build his acceptance speech around all the little quips he had gathered from the guests he'd spoken to beforehand. It was remarkable how quickly he developed a stand-up comedy routine. As anyone who has seen him perform knows, Williams was an improv genius, and that was fully on display during his tribute.

The evening featured video tributes from an assortment of Robin's friends in the business, including Billy Crystal, Whoopi Goldberg, and Harold Ramis. I presented him with our Gold Hugo career achievement award, which he kept stroking like a giant dildo. He said it reminded him "of a melting Oscar® statue," which he followed with a 45-minute acceptance speech that seemed to cover a little bit of everything (including Chicago corruption).

"Here we are sitting near the new Millennium Park, which is already a decade unfinished and $300 million overbudget, but that's city hall," Robin pointed out. "We were very religious in Lake Forest. My mother was a Christian *Dior* Scientist." He seemed to be clean off of drugs and alcohol that night, which was naturally a relief. He mentioned that his own kids were up in his hotel room watching porn and having a couple of cocktails.

"The kids keep asking, 'Dad, did you do a film *called Good Year Humping*? And another one called *Romancing the Bone*?'"

His people made it clear that we must keep our wine or any alcohol sponsor away from him at the table or near the stage, which was funny since the first thing he did on stage during the speech was grab a bottle of wine from a table and pretend to drink the whole bottle,"

It was Robin's 52nd birthday that day, and we had a four-foot-tall layer cake to surprise him. I think he was touched.

--

Of all the film directors who have attended our Chicago Festival over the years, French directors topped the list. We had Francois Truffaut, Bertrand Tavernier, Patrice Chereau, Agnes Varda, Jacques Demy, Leos Carax, Jacques Tati, Francois Ozon, and Luc Besson. But it was Claude Lelouch who appeared the most. Why? Well, he tended to make a new film every two years, and I liked his work, as did Chicago audiences. And he liked to come to Chicago.

My relationship with Lelouch extended back to 1966, when we were in Cannes at the festival. I was attending with Colleen Moore and King Vidor. King was on the jury and asked each day if I had seen anything that I would recommend to him.

It was during that festival that I saw an unusual, dazzling, kinetic style film that I couldn't stop thinking of. It was *A Man and a Woman*, set to a Francis Lai score, with quick editing and flashy camera work, besides being very sophisticated and tender in the romance department. You had an irresistible chemistry between the two leads, Anouk Aimee and Jean-Louis Trintignant. It was different from everything else that I had seen so far, and dazzling.

Lelouch had made a few films before this one. He seemed to be a known commodity in France, but not to me. He was a director, writer, cinematographer, actor, and producer, one who was always taking chances and delivering a lush, romantic, visual style, scored with memorable music. You either loved his films or hated them. Audiences loved his style. Film critics? Not so much.

I remember when we did a tribute to Lelouch years ago in Chicago, screening a half-dozen of his films. This kind of honor for him was frowned upon by the local press, but I knew it would sell out. Claude always came to our festival with a new film and often a new wife. He had five wives and seven children in all.

One year, his son Sachka Lelouch came with him, as he was the cameraman on Claude's latest film. We were sitting together one afternoon when he said, "You probably knew my mom."

"Well, I probably do," I replied memorably, "but which one was she?"

A stunning fact of Claude's childhood was the fact that he was hidden from the Gestapo in movie houses and saw endless films. He always said, "The cinema saved my life." Claude is now in his eighties, has made around fifty films, and is harder at work than ever. He has a wonderful headquarters in Paris, with offices, a fabulous screening room, and a private dining club and restaurant. Not a bad way to live and work.

I recall Claude telling me a story about the time he hosted a private screening for Charlie Chaplin at his place. After the screening, Charlie tipped Claude, thinking he was the usher at the theater. Great story. I love the guy.

--

Through my festival, I have been privileged to meet so many of the greats, from Chaplin to Orson Welles to Spielberg and others too numerous to mention (though I'm sure I'll inevitably mention them). But it was a Chicago genius who became family to me.

I read about Ken Nordine in the newspaper as the creator of something called Word Jazz. I initially had no idea what that was, but I found out quickly enough by hearing it every night at midnight on FM radio. This man spoke poetry, which typically doesn't thrill me, to electronic sounds and melodies that he would create in a hypnotic, surreal form that soothed you and excited your imagination at the same time.

Ken was very special. He was brilliant, actually. He had a vision, and he could technically create it all by himself. He could see art in full color right before his eyes. He made a short entitled *Black* that I saw on PBS and invited him to the festival in 1966. That's how we met. I wanted to know more about him and went to his studio, where he offered to work with me and the festival doing some creative new ideas.

Remember, I was a mere child of 23 at this point, and Ken made me feel like I was family. He had a booming voice that sounded like God speaking from above, deep and powerful. I learned he made his big bucks doing voiceovers for TV commercials. Automobiles, coffee, blue jeans, you name it, he was hawking it. I heard him all the time and didn't even realize it was him.

Ken was so popular with the ad agency guys that they came to *him*, so he never had to shlep to New York or Los Angeles. He built this sophisticated studio atop his old mansion in the Edgewater neighborhood of Chicago, just down the street from Loyola University. He was known as "The Voice" and, appropriately, became the voice of the Chicago International Film Festival for fifty years.

He has one of the most distinctive voices in the aural landscape, and we were lucky to have Ken as our man for so long.

10 – THE WOMEN

The Chicago Festival was being covered in 1980 by a young journalist from London named Jan Dawson. The first thing she attended was a cocktail party for Neil Simon, whom we were honoring and showing his latest film, *Chapter Two*, starring his new wife, Marsha Mason.

Jan's first observation wasn't Neil or master of ceremonies Rex Reed. It was the curious first names of the ladies at the party. Her article back in London was headlined *Sugar, Cookie, Bootsie, and Pussy Star at Chicago Festival.*

When you work in the arts anywhere in the world, you depend on many socially connected people. Mine just had some cool names. "Sugar" Donna Rautbord was a society/PR friend. Anida Johnson "Cookie" Cohen was married to an automotive millionaire at a time when being a millionaire was still a big deal. "Bootsie" was philanthropist Elizabeth Nathan. No clue where the Bootsie came from.

Then we had "Pussy," Mrs. Walter Paepcke. Her husband was an industrialist and philanthropist who created Aspen and the arts in that mountain skiing community of millionaires. I never dared ask where her nickname came from.

When it came to events honoring woman, we had some great ones. None was more entertaining than our evening with Liza Minnelli during the festival in 2006. Liza loved Chicago and agreed to the tribute with the help of Sugar Rautbord. I thought it would be perfect to have British film critic John Russell Taylor interview her on stage as John had interviewed her father Vicente Minnelli in 1974 at the festival.

However, I knew there would be some problems from the start, owing to the fact that our official hotel didn't allow dogs or smoking (the smoking ban being citywide in all Chicago hotels).

Liza arrived in good spirits despite being in the middle of her divorce from David Gest, who was suing her for $10 million while claiming she was violent and physically abusive due to her persistent alcoholism. Poor Liza.

Ms. Minnelli was in great form, however, doing press on the red carpet until one local critic asked foolishly about her current divorce situation. That ended the press interviews. The audience was a full house. Seven-hundred and fifty fans of Liza jammed in. Young, old, straight, gay, all were there, and they brought along flowers and plenty of love.

The old queens in the audience saw Liza as Judy Garland's daughter. The young ones knew her as a Broadway and concert star and Academy Award® winner for *Cabaret*.

John and Liza comfortably carried the evening with film clips and a slew of wonderful memories both about her mother Judy and her father, the great stage and film director Vincente Minnelli. Liza came with a manager/assistant who was basically a pain in the ass, seemingly along only to make things as difficult as possible. All he could ever say was, "No, she will not do that! No! No! No!" But Liza herself always said yes. She lived up to everything we had agreed to at the start. She even sang an impromptu song on stage, which pleased her fans no end.

During the interview, Liza turned from John for a moment and somehow had a lit cigarette in her hand. She continued to smoke while talking. We knew she was a chain smoker, but how she actually managed to slip the cigarette lighting part of it past us was a mystery.

Meanwhile, we thought we had covered all our bases. But no one remembered that Liza had just had a hip and knee replaced and couldn't even attempt the flight of stairs to get to the afterparty. We found the freight elevator and lifted her into it from the ground level. She naturally had to keep smoking during this ordeal.

Anyway, the elevator took her to the kitchen of the Blackbird Restaurant, where the party was taking place. The chefs were freaking out with all the cigarette smoke wafting onto their food and diners. Liza entered the party and greeted the seventy-five or so guests. The restaurant wouldn't bend on the smoking situation, so Liza declared, "I'm outta here!"

Back into the freight elevator she went. Liza was a real trouper. Her assistant was dumped the following week, I'm pleased to report.

--

I love Norwegian actress Liv Ullmann. We met at a dinner party given by the Swiss producer Arthur Cohn in Cannes, and we remained close friends from that moment on.

Liv was one of many muses that the famed Swedish film director Ingmar Bergman cast in his films. Bergman married several of his actresses, but Liv remained both an actress and sex partner, even having a child with Bergman, but never marrying him. She did eight remarkable films with him, including *Cries & Whispers* (1972), *Hour of the Wolf* (1968), and *The Passion of Anna* (1969).

But I feel that *Persona* (1966) was the movie that turned Ullman's career around and made her an international star.

Back in 1990, I did a tribute to Bergman in Chicago and invited all of his ladies to be part of the tribute along with The Master himself. The ladies came; Bergman didn't. Having Ingrid Thulin, Bibi Andersson, Harriet Andersson, and Liv Ullmann all together in the same location was spectacular. They all seemed to like one another, which I found surprising considering all had been sexual partners with him.

Liv was about to direct her first feature film, and I asked if we could premiere it whenever it was completed. She said, "Of course, I would be honored, and I will be there for it."

A few years later, in 1992, Liv arrived for the premiere of *Sofie* at our festival. We used a vintage movie house, Music Box Theatre. It had 750 seats, perfect for the film. Liv was on stage introducing her first full-length directorial feature. The audience was packed and thrilled to be there.

The drama was about to begin.

Ullman left the stage and came out to the car where I was waiting with her husband, Donald Saunders. Suddenly, the theater manager came sprinting out to the car yelling that the film had the wrong subtitles. It seemed the film had come straight from the airport and had not been checked by anyone at the festival or the theater.

"NOOOOOOOO!!" Liv wailed from the car.

Liv scurried back into the theater, down the aisle and onto the stage, crying and apologizing to the audience.

"I will stay and translate it to you," she vowed, pleading, "Please stay and I promise that we will have a new print by tomorrow." The audience stayed. She came back to the car hysterical.

"Let's go to the dinner and calm down," I said.

"I just want to go back to the hotel," Liv replied.

I suggested we have a drink at the dinner, then go back. She agreed. At the dinner, I thought the best thing was to get her a vodka and then another vodka and then another (and maybe a fourth?) to help forget about this situation and stop crying.

Alcohol. The answer to all our problems.

It worked. However, it brought out another, uh, minor issue. Liv turned to her husband and said with chilling simplicity. "I hate you and I want a divorce." Right in front of me!

Um, what?

Suddenly, the problem with the subtitles seemed perfectly manageable.

I paused for a few seconds after hearing this and finally said, "Look, let's solve the film problem first, then you two can work the other thing out later."

We found the proper print, had it shipped in, and rescheduled the film for a few days later. Liv was already on a flight out with her husband.

I was with Liv in Los Angeles in 2022 at yet another dinner party for Arthur Cohn, his 95th birthday. Liv said, "You know, Michael, I have never had another vodka since that night back in '92, and yes, I am still with Donald." She did have the person who made the mistake with the film print fired from the Film Institute. What goes around, comes around. You can't go pissing off film legends and get away with it.

--

I need to say a few words about Angela Lansbury. From the big screen to the Broadway stage, who was more fabulous than Angela?

In the beginning, she was a contract player at MGM, doing lesser roles in *Gaslight* (1944) and *The Picture of Dorian Gray* (1945) and was particularly noticed in '62 in *The Manchurian Candidate*. But she didn't become a full-fledged star with her name above the title until she hit the stage in New York in 1966 with *Mame*. After that, she owned Broadway, and we knew we needed to honor her.

I had never met Angela. I knew that Rex Reed, the New York entertainment columnist, and she were best friends. He had done on-stage interviews with us at the festival for Ann-Margret and Neil Simon and would be perfect with Angela, we decided. Rex would speak to her on stage after we screened a new film she had just completed for theatre director Hal Prince, *Something for Everyone*. It was gonna be a very exciting evening. We just knew it.

Indeed, it was a sellout, 1,400 seats at the old art deco Esquire Theatre on Oak Street. And yes, it was a young gay "We Love You Angela!" audience in attendance. Angela had become a gay icon because of *Mame*. As for *Something for Everyone*, it wasn't great, but it did show off a young Michael York, who would soon hit the big time for his bisexual role in 1972's *Cabaret*. The story is reminiscent of Paolo Pasolini's *Theorem,* featuring another young man who would sleep with anyone to get to the top. In the Pasolini film, it was Terence Stamp as a Christlike figure. Here, it's a cute young guy whom the cameraman seemed to fall in love with, played by York.

Angela was a bit too young for the part of an aging penniless countess. She was playing somewhere between her *Mame* character and her Tony-winning *Dear World* role (Broadway, 1969), but she managed eventually to pull it off. She and Rex wowed the Chicago Festival crowd. Then we went on to the afterparty, which was near the top of the nearby 100-story John Hancock Building, the tallest building in Chicago at the time.

It was a stormy night, which didn't concern me till Angela and I reached the 96th floor festivities. I had never considered what the building might be like in a strong wind. The sway was noticeable, and Angela walked into the party, said hello to the key people, then came to me and said, "Get me out of here, my ears, my balance, I can't be here." Out we went.

Angela left show business for some years to take care of family matters back in Ireland with her drug-addicted son, only to return in 1973 to star in *Gypsy*, the role that Ethel Merman made famous on Broadway. The new production played London, then went on an American tour before Broadway.

When the national tour hit Chicago, I met Angela once again. This time, I thought maybe we should shoot her for a film festival commemorative poster, as we had done with so many stars. I took her over to meet our photographer extraordinaire Victor Skrebneski. He was gracious as always, but something didn't click with Angela.

Victor took her on a tour of the studio and showed her his previous work. Angela pulled me aside and once again demanded, as she had so many years before, "Get me out of here!". We left, and on the way back to the hotel she said she felt he was "demonic in some way." That's what you call not connecting.

At dinner that night, just the two of us, Angela surprised me with a curious question: "Well, you boys have your bathhouses and such, but what about me?"

I felt she was asking me to arrange some sort of on-the-road sexual adventure with an escort. I was taken aback, and didn't address the request, instead changing the subject to, "How can you do this on the road, eight shows a week, for years at a time? I don't understand," Angela replied. "The moment the orchestra plays the first chord, I'm out there giving every ounce of me to that audience. I am dedicated and expect my entire cast to have that same drive."

She was clearly a perfectionist. I ordered us a few more drinks.

It was 1980 before I would be with Angela again in Chicago, when she was starring in the national tour of *Sweeney Todd*. I took her and her husband Peter Shaw, the film executive, along with hairstylist and superfan John Lanzendorf out for dinner a few days before opening night.

I had met Peter in California years before. He was a delightful guy and a loving husband, even if he was known to play both sides of the fence. When Angela went to the ladies' room, Peter asked, "Where is the nearest bathhouse?"

Nothing much surprised me anymore. But why did I always feel like the pimp?

--

I'd been trying to honor Shirley MacLaine in Chicago for more than a decade. Her publicist Dale Olson had always said NO. He was also handling Rock Hudson in his last days, having to deny that Rock was dying of AIDS. Why Dale said YES in 2005, I will never know.

Shirley required a private jet, and not just a regular one but a 727, which was popular for commercial transport back then. It held 106 passengers. On this particular flight, there was only Shirley. Thank God I had a Chicago Festival board member, Terry Schwartz, with a generous father who came up with the $25,000 to fly her to Chicago.

I met Shirley at the private airport to bring her to her hotel. Along the way, she wanted to telephone her home in Santa Fe, New Mexico, to speak to her dog. (Remember, she channels people and animals from past and present.) At the time, I didn't think it was odd at all.

Our driver had a heavy foot, so every so often he would accelerate on the gas pedal and brake on and off.

"If you keep driving like this," Shirley warned, "I will vomit on you."

The driver calmed down on his manner of driving. I asked Shirley if she would do a photo shot for us as a poster by our famous photographer Skrebneski. I showed her a book of his work.

"He's gay, right?" she asked.

"Well," I replied, "you'll have to discuss that with him."

The evening of the tribute arrived. Shirley looked terrific in her Vera Wang gown and sat at the main table with actor-writer-director Harold Ramis and some fellow board members, including the father who had paid for the jet. The tribute started with Roger Ebert and video tributes from Sally Field and others.

When it came to her acceptance speech, Shirley was gracious and personal, and then very naughty. She asked the audience if they wanted to know about moments from some of her films. That's when the fun began.

Said she: "When I watched (highlights from) my fifty years on the screen tonight, I was so glad you didn't show the masturbation scene from *Being There* with Peter Sellers. When I was working with director Billy Wilder and Jack Lemmon doing *The Apartment*, Billy believed that for us to better understand the story, Jack and I should visit a whorehouse and watch how the girls did their job through a peephole for two days. I think that's why I had so many love affairs from that moment on in my life."

Shirley continued, "I loved working with Clint Eastwood, but living next-door to him on the golf course back then and all those Swedish airline stewardesses, oh my God!"

When it came to channeling, she said, "Nobody does it better than Meryl Streep. We did *Postcards from the Edge,* and Meryl played my daughter. She was remarkable becoming that person. It was an honor to work with her."

Then there was *Terms of Endearment* with Debra Winger.

"If you remember the film," Shirley said, "Jack Nicholson and I were in bed together. Unbeknownst to the film's director and the audience, Debra was under the covers causing chaos just for the hell of it. She was going UP on both of us with her tongue! Now I'm a pretty good sport, being a dancer and all that team-playing stuff, but this was going too far.

"Jack said, 'Ya know, don't knock it if ya haven't tried it.' I punched him in the balls and said, 'I'm going to tell my brother (Warren Beatty), and you just try it with him!'"

Shirley concluded her thoughts with this: "I am proud to be in this business of show, and I understand the business part, too. Commitment is important to allow creativity to happen. Learn all you can about self and that film is more lasting than politics."

--

In 1992, we initiated the Piper-Heidsieck Champagne Award. Why? Because as a new sponsor, they gave the festival $25,000 and promised the honoree a case of Piper-Heidsieck bottles for a couple of years. That's why.

As actress Kathleen Turner was really hot in so many terrific films at the time, we wanted her as the first Piper-Heidsieck honoree. Her roles included work in *Body Heat, Romancing the Stone, Prizzi's Honor, The War of the Roses, Peggy Sue Got Married*, and something that still makes me laugh out loud, *Serial Mom*. She was sexy and a great actress who could handle both comedy and drama smoothly. She was one of a kind, with that famous gravelly, husky voice.

We did the tribute. I think she had fun with it. First came a clip reel, followed by an interview on stage with famed London film critic John Russell Taylor. Despite his rapturous introduction, Kathleen actually was able to get a few words in. John loves to go into lots of detail about the honoree and forgets they're sitting right there. She mentioned that people still ask her to recite the famous Jessica Rabbit catchphrase, "I'm not bad, I'm just drawn that way" from *Who Framed Roger Rabbit?*

Kathleen was a professional through and through, in striking contrast to the stories we would read at that time about her temper and diva-like fights with directors and cast members. I suspect that she made enemies in Hollywood because she did a lot to try to correct some things lacking in Hollywood's attitude surrounding women's rights, salaries, and basic respect.

Remember that back then, women were widely thought to have no right to stand up to men. If things have changed today, Kathleen has no doubt had a lot to do with that.

Kathleen came back to Chicago to be the president of our feature film jury one year, and did a second evening in person, as frank and disarmingly honest as the first. Audiences love to hear her stories. Me too.

I spent time with her in Chicago when she was doing her one-woman cabaret show entitled *Finding My Voice*. Who knew she could sing? She can, and she can tell stories about her life, her drinking, her pain, and her difficult existence with rheumatoid arthritis, which kept her out of the business for a while.

She emerged back in the spotlight with Michael Douglas on the Netflix series *The Kominsky Method*, about an aging acting coach who runs an acting school. Kathleen plays his third ex-wife on the series. Kathleen and Michael made three films together, all hits.

Of course, I had to ask the question. No, she never had sex with Douglas, but they came close, she says, until his first ex-wife would show up on the set and say, "Remember, he's mine!".

--

I always wanted to honor French actress Catherine Deneuve, but I could never interest my board members that she was important enough. The members had to raise money by selling the gala tables at benefits, and they also had to love the honoree.

Deneuve is a fantastic actress, model, and producer. She's considered one of the greatest European actresses ever, appearing in everything from *The Umbrellas of Cherbourg*, *Belle de Jour*, *The Truth*, and *The Young Girls of Rochefort*, to *Indochine*. We've shown one of her films in many editions of the Chicago Festival, always to sellouts. While I never found Deneuve fabulously sexy like Sophia Loren, I thought her handsome and really beautiful at times.

I still find Catherine captivating on the screen today. She's aged well. Even though I couldn't convince my group to buy into her as an honoree, I thought I would find out what it would take to get her to Chicago, maybe for a film tribute or a master class. Since UniFrance, our French film contacts, knew her agent and manager, I asked them to get me some details.

At that time, the conditions were these: two first class tickets on the Air France Concorde from Paris to New York, then on to Chicago. Already, that was too pricey. Her people said it was important for her to spend time in New York, as she loves to shop. Basically, they weren't interested in a tribute or an appearance in Chicago.

Years later, I checked with other festivals and found that Berlin had paid Deneuve $75,000 for an appearance, and another festival in Romania had matched that. I never wanted our festival to open that Pandora's Box of paying for talent. Once you start, it doesn't end. So, Catherine never made it to our festival, but she *did* make it to Chicago.

In 1971, I tossed a cocktail party at my place in Old Town for the Italian screenwriter and director Ettore Scola, who was shooting a film here. *My Name Is Rocco Papale* starring Marcello Mastroianni was using a location in a bar right under the film festival office. I talked to Marcello, who was young and friendly. I was a great fan of his from the Fellini films.

What I found out was that Marcello was staying at the hotel in Chicago called the Ambassador East, which is famous for having The Pump Room restaurant. Across the street was the Churchill Hotel, and Marcello was having a secret affair with a beautiful woman staying at the Churchill. Who knew that the woman was Catherine Deneuve? I certainly didn't. Her hair in a red wig, she was young and beautiful, wore giant dark sunglasses, and was not at all recognizable. They conceived a child in Chicago. That child turned out to be actress/singer Chiara Mastroianni.

--

It was in 1992, during all the controversy surrounding a new film on the JFK assassination from director Oliver Stone. that I met Paula Wagner. I wanted to honor Stone and also do a high-profile live panel with Oliver, famed Chicago author and historian Studs Terkel, Roger Ebert, and a few experts on the various assassination theories. Paula was Oliver's agent as well as a film and Broadway producer. She made the event happen.

The gala, award and panel were a sensation. I would meet Paula again when we honored Tom Cruise a year later. We came up with a hook: Actor of The Decade, as Tom had just made his tenth film and had some history with Chicago, having made *Endless Love* and his breakthrough film *Risky Business* here. When Tom came for the gala that year, he was THE hot property. You could tell because he was surrounded by the two power-broker super agents of CAA, Mike Ovitz and Ron Meyer, along with PR terror Pat Kingsley. Also along for that ride were Tom's mother and stepfather, his then-wife Nicole Kidman, and Paula, who was managing Tom at CAA.

Paula was this young, feisty agent, handler, perfectionist and rep for Tom. Together they would later develop Cruise/Wagner Productions to make the first three *Mission: Impossible* films along with one of my favorite movies, *Shattered Glass*. Paula eventually went out on her own as an executive producer of the Steven Spielberg flicks *Minority Report* and *War of the Worlds* and later *Marshall*, a biopic about Supreme Court Justice Thurgood Marshall starring Chadwick Boseman. She is a dynamo with everything she touches, including the musical stage production of *Pretty Woman*.

Paula has been a mentor to me, always believing in what I was striving to accomplish with the Chicago Festival, from discovering new directors to honoring the past and present of the business. I recall sitting in her office at United Artists, which she and Tom had committed to revive for a few years, or in the office of her husband, entertainment executive Rick Nicita. We would all brainstorm ideas and talent.

Paula and Rick knew everyone in the business, even my producer Arthur Cohn back in Switzerland. They could just pick up the phone and call publicist Stan Rosenfield or lawyer Barry Hirsch and make things happen. They were movers and shakers before the term was even invented.

--

I never thought I would meet Charlie Chaplin's daughter, Geraldine. And if I had, who would have thought it would be in Havana, Cuba? Of course, I had seen *Dr. Zhivago*, and many Spanish films that she acted in for her then-partner, director Carlos Saura, but I was transfixed with her in a Claude Lelouch film, *Les Uns et Les Autres* (*Bolero* in the USA). Remember, I'm a giant Lelouch fan. Then of course, she was also in *Nashville* and *Welcome to L.A.* and *The Impossible*.

I introduced myself to Geraldine as a friend of the writer of the film *Chaplin*, David Robinson. We spent ten days together at the Havana Film Festival in 2015 watching international films from Latin America with her famed Chilean cinematographer husband, Patricio Castilla. What a riot he was. They also have a very talented daughter, filmmaker Oona Chaplin, who hated me right away because we turned her film down in Chicago a while back. Like I knew? Anyway, Geraldine and Pato knew Havana well, and driving around between films was an adventure.

The theater where we screened our films every day was dedicated to Geraldine's father: the Chaplin Theater. I had stayed a few times in Havana at the famed Hotel Nacional de Cuba. The place had history with gangster Lucky Luciano and then the 1950s gambling days of another organized crime figure, Meyer Lansky, and outgrossed even Las Vegas in its day before Castro took over in Cuba in 1959.

All that was gone by the time I got to Havana. Yet it remained a tourist draw.

We and all film festival guests stayed and played at the fabled hotel. The casino was now the breakfast room. There was still a splashy Lido-style showroom of sorts in the lower level. The swimming pool was exactly as it was. It had elements reminiscent of the Royal Hawaiian Hotel and the Beverly Hills Hotel, and why not? They all played their part in the design of the place.

But back to Geraldine and Pato. I have never met such a devoted couple. He toted around a toaster, making toast for her and serving her breakfast in bed every morning. She's a down-to-earth person, greeting guests who wanted to meet her in the hotel lobby and even in the street. She was easily recognizable as...someone. Why? Cause she has that Charlie Chaplin face and presence.

I watched strangers approach Geraldine and act as if they knew her. They had no idea who she was, but it didn't matter. I asked her one day if it bothered her.

"No," she replied. And she meant it. Geraldine was happy to meet anyone who wanted to meet her.

We spoke at length about her legendary father, working with Richard Attenborough, and spending time with Robert Downey Jr., since she played his mother in the 1992 film *Chaplin*. "I never really ever interacted with him," she admitted. "He came in and did the scene and then was gone."

After Cuba, I invited Geraldine and Pato to Chicago to be on two separate film juries at the festival. I also thought a tribute to her film career would be special and that we should do it at the Essanay Film Studio, where her father made the short film *My New Job* in 1915 and we did the first Chicago Film Festival awards gala in 1965.

It was magical, and I think Geraldine really felt it that night with all her fellow performers sending in video tributes to her. She said, "I feel my dad's spirit is here. There are so many ghosts. Gloria Swanson, Ben Turpin, Wallace Beery."

It was amazing that Claude Lelouch was in the audience, as we had presented the premiere of his latest film a few days earlier. It was a remarkable reunion, since they'd worked together back in 1981. I recall Lelouch saying to Geraldine, "We should consider working together again." Everything came full circle that night.

It's a challenge to follow the supreme achievement of a universal artist like Chaplin, but Geraldine's seven siblings have also made their own lives. Michael had his moment of stardom as a boy, in *A King in New York*, but found farming and quiet charitable works more rewarding. Josephine, twice widowed, had a successful career as an actor in the '70's and '80's. Victoria – who Chaplin said was the one who had inherited the gift of comedy – has toured for decades with her husband Jean-Baptiste Thierrée in their magical two-person circus. Their children, Aurelia and James, have each created their own shows, guided and designed by Victoria herself.

Eugene has remained in Switzerland, experimenting with various business interests. Jane was briefly married to Ilya Salkind, during which time she partly financed his *Christopher Columbus: The Discovery*, but now lives in Latin America where she raised her two children. Annette successfully tours with her own theatre company. Christopher, the youngest, had a brief career as an actor before dedicating himself to music and finding success as a composer. He also enjoys piloting his own plane.

The Chaplins are, to say the least, versatile.

I mean, imagine being the child of Charlie Chaplin. Wow.

11 – SORRY, WRONG NUMBER

I've been thinking about many of the personal requests over the years from directors, stars, and visiting personalities attending the Chicago Festival. I was always shocked, as I held them all in such high esteem, and of course didn't really know any of them personally until the moment they arrived in town. And once in town, they felt like they could misbehave, because they were technically away from their everyday lives.

I never considered that the stars and top filmmakers were only human and suddenly had human desires, and as their host, I was somehow expected to fulfill them. Who knew this was the responsibility of a festival founder?

Take producer Allan Carr. He was a fellow Chicagoan and a big deal due to the movie musical *Grease*. He gave us the world premiere in 1980 of *Can't Stop The Music*, a God-awful film, but with a fun campy cast like the Village People and Bruce Jenner (the gold medal Olympic decathlon champ who has since become the transgender woman Caitlyn Jenner).

The Village People were *the* hot gay musical group at the time with the trademark song "Y.M.C.A." Alan decided that he wanted a 17-year-old blue-eyed, muscular blond in a red Speedo bathing suit in his hotel room, stat.

"Sorry Alan, we can't help you," I assured him, "but I'm sure the PR firm handling the film can fill your request." It was probably easy, as the film premiere had ten of these young hunks, all as props at the after-party.

--

Everybody who knows anything about director Abel Ferrara (*Bad Lieutenant*) will know about his heroin addiction. At least in Chicago this time, he only needed a fix for a cast member.

Peter Kern, a delightful, crazed Austrian actor/director (*Fox and his Friends* and many Rainer Werner Fassbinder films) loved very young men and demanded a 16-year-old boy in his hotel room in an hour, and another the day after that as well. My staff was shocked. I kept wondering, "What can a 350-pound man do with a kid except crush him?"

No, we did not fulfill his request. That would have been just plain wrong.

Then there was famed director George Cukor (director of the 1954 version of *A Star is Born* with Judy Garland and 1964's *My Fair Lady*). I would take him out to dinner during the festival, and he had his usual requests. Once, he had the hots for a waiter who caught his eye in the restaurant.

"I want to suck his cock," he announced, apropos of nothing.

"George, I'm afraid we can't arrange that," I assured him. "You are our guest here, and you have to behave."

"Why?" he asked, incredulous. "I want what I want, when I want it. And I want it *now*."

As Mick Jagger once so famously sang, you can't always get what you want.

One of my favorite moments of festival misbehavior involved actress Shelley Winters (*The Poseidon Adventure, The Diary of Anne Frank, A Place in the Sun*), which should surprise no one who knew her. She had been a showstopping beauty once upon a time. I always loved her films and was excited to meet her.

We were presenting Shelley with an award on stage at the prestigious Chicago Theatre. We had the Mayor, Richard M. Daley, present the award and a key to the city. Shelley was delighted and said to him, "It is so good to see you again after all these years. How do you do it? You look like a 50-year-old! I want the name of your doctor."

Of course, she thought he was Mayor Richard *J.* Daley, his father, whom she had known way back in the day and who was also in charge of Chicago back then. It was a riot.

I took Shelley out after the theater to my favorite hang-out restaurant, Kelly Mondelli's. After her film that night, she was feeling no pain, and a few drinks later she said to me, "See that guy at the bar?"

"The truck-driver sort-of-guy?" I asked.

"Yes. I want to fuck him!" she said.

I was floored, but asked the restaurant owner Joey Mondelli if he would ask the guy if he would like to meet this actress. I have no idea how that ended.

--

In the festival office, I would learn years later that French director Jacques Demy (*The Umbrellas of Cherbourg*) was carrying on with a young man who was my intern and was the projectionist in our office. His wife Agnes Varda didn't know and evidently didn't care at that point, as they had separated.

After Demy's death, Agnes made a number of affectionate film tributes to her husband. Critic and author John Russell Taylor, ever ready with an appropriate quote from English literature, said that Jacques's attitude toward attractive young men reminded him of Alexander Pope's description of a dog "mumbling the game he dared not bite." I learned that my staff applauded the day she left Chicago. I guess she had a lot of demands. Nonetheless, I liked Agnes. We met many times after Chicago in Cannes, and by then she was much more mellow.

Demy was not the only visitor to find relief in the arms of a festival employee. During his visit to Chicago, I subsequently learned that Indian director Satyajit Ray (why was I usually the last one to know?) was having a two-night affair with a festival staff member. At least that time, it was some-one of the opposite sex. Not that it mattered.

--

The legendary silent film comedy star and stunt performer Harold Lloyd (*Safety Last, The Freshman*, and many other great silents) asked if I could send over that young college girl he spotted in the audience to his hotel room. Mind you, the man was in his 70's at this point.

"Harold, you are a dirty old man!" I chided.

"Yes," he admitted benignly.

I don't think the college girl bit.

--

Frank Ripploh, the German actor/director of *Taxi zum Klo*, a gay film that itself was plenty controversial, arrived, checked into his hotel, then vanished completely for three or four days, without a word to anybody. He turned up only on the day his film was being shown, blandly explaining he had checked into the nearest YMCA for a few days of sexual fun – a desire apparently satisfied beyond his wildest hopes.

Then there was the experience of the prize-winning English student film-maker who got lightly mugged walking across Lincoln Park near his hotel. He said he couldn't understand why this big black guy kept calling him his mother. We worked out that what the mugger was actually saying was, "Hand over your money, you mother," a usage presumably not familiar in Oxford or Cambridge.

--

How many of my married board members over the years wanted to borrow my apartment for an afternoon liaison? Let's just say it was a lot. I was like Jack Lemmon in *The Apartment*. I recall saying, "I have bunk beds with Mickey Mouse sheets on them. If you two can handle that, I guess I can give you the keys."

Does that make me a sexual go-between? I think it does. And I guess I had to be fine with that.

--

A six-foot-four, 28-year-old European director fell for one of our young hospitality ladies. They were all basically interns or temporary staff and of course delighted to meet the stars and directors. They would do most anything, it seemed, for our guests. So, these two hit it off. The problem, it turned out, was that the poor fellow didn't perform well. Sad, because she then told the rest of the team, "Nice guy, small dick."

I couldn't look at the guy after that when I saw him in Cannes without thinking, "Nice guy, small dick."

--

The Russian and Polish directors who came to our festival had simple requests. They were looking for prostitutes and the location of the sex shops for porn. When it came to Russian vodka, fortunately, we had plenty of it, as our sponsor that year was Stolichnaya.

Back in the Sixties, the National Film Board of Canada sent us one of its finest filmmakers. We were thrilled to have him at the Chicago Festival until he disappeared for two days. Just flat-out vanished. If only they had warned us that he was a total alcoholic, at least we would have known where to start looking. Instead, we are calling hospitals and police stations. He miraculously reappeared the evening of his tribute with no memory of anything being amiss. And he was brilliant on stage.

An Italian film critic on our feature film jury asked if he could stay an extra week after the festival had ended in our official hotel because he had fallen head-over-heels in love with this other, female, festival jury member.

"Sure, but aren't you gay?" I asked innocently.

"Yes, I was," he replied, "but now I'm not."

The iconic French actress/singer/director Jeanne Moreau (*Jules and Jim*) was at the Chicago Festival several times when she was younger and hot. She came once with a feature film she was starring in, and a second time as part of a film jury. She always had her own driver. They were usually in their early twenties. I didn't expect her to fall for one of them.

One day, Jeanne called me to the hotel and ordered, "Fire that driver!" I didn't ask the reason, because I was too impressed with Jeanne to question it. Of course, when I asked the driver, he told me they had been having a slight fling, and he wanted out, as she was too old for him. She was pissed and wanted a new driver. Understandable, right? "Give the lady what she wants" was a famous motto of Marshall Field's department store here in Chicago back in the day. So, Jeanne got her new driver.

--

If you ever had a chance to meet actor/producer/director Maximilian Schell, you could understand immediately that whatever he wanted he would get from any woman he met at any festival. He was a romantic, charismatic Austrian actor, instantly recognizable and always horny, and women were utterly drawn to him. I watched it as two different board members fell under his spell. They both got what they wanted. I didn't have to do a thing.

12 – THE OMEN

The American concept of a board of directors to manage artistic and cultural entities is unique. In Europe, of course, film festivals or theatre companies have their boards, but these are largely made up of people with special contacts or particular expertise to help lead the administration in question and who are often paid appropriately for their service.

To the contrary, in most instances in the United States, the members of the board are actually private sponsors, generously contributing to the running costs of the organization. These contributions are mostly altruistic and voluntary, people simply and selflessly promoting a particular favorite art. There are inevitably small incidental paybacks involving social access to the great names of the art, and the pride of seeing your own name figuratively in lights.

Occasional enthusiasts enjoy flaunting their multiple board memberships like jewels that proclaim their patronage of the national culture. But whatever the motive, a good board can offer valuable advice and governance to the director and staff.

So, we were excited when we lured Ellis Goodman, celebrated for his sponsorship and encouragement of entertainment arts In Chicago, to join the Chicago International Film Festival Board of Directors.

Goodman was an English-born businessman who started his career as an accountant with prestigious clients who included singer Petula Clark but moved to the United States to build up an awesome financial empire established on beer and the merging of brewery companies. As a board member, he was not only very shrewd but also arranged sponsorships from his beer concerns.

We all yearned to have Goodman as our chairman, but he was never available because his business kept him constantly traveling and negotiating over his interests in Europe, Mexico and even – that far back – China. So, while his company grew all the time, he was occupied traveling to these far-off places taking over companies and firing people, which was part of the process.

The man was very hands-on in business. I remember him asking me once rather anxiously if I thought the breasts of the girl on the label of St. Pauli Girl beer were impressive enough. (Their slogan: "You never forget your first girl.") So, he didn't have time to be fully on board with us, but he would pop in and give us sponsorship.

And then one year, Goodman arrived and announced, "Well, I think I have time now. I've done my thing. And now I'll be your chairman." I think at the time, he had chosen between us and the famous Steppenwolf Theatre. Or maybe the Steppenwolf wasn't ready to vote him in as chairman, which disappointed him. He would always say at that time, "No one knows me in Chicago." But that was about to change.

Goodman was energetic and stepped right into things. He said, "Okay, you need a person running this whole thing, so I'll get you... someone." He brought in three or four people for me to interview, and I agreed on the one whom it was later clear he already wanted to take the title of managing director. This was Elizabeth Morris, a.k.a. Betsy, who died recently at 63 after a life largely devoted to philanthropic and artistic causes. When she came to us, she was young and irresistibly sweet and charming.

Ellis's next startling statement was, "But all of you people are working for nothing. We have to work out a new salary structure and make the thing better for everybody." So, endless meetings ensued. There was a process. You went to his offices. The secretary would bring you in and he would sit you down and you'd be eye to eye, under his control.

"Here's what I'm going to do and here's what I'm willing to do," he'd say. I instantly trusted this man, which was something rare for me. I decided this was going to be great. I was enthused.

"Well, the first thing we have to do, Ellis, is to take you to the Cannes Film Festival to see how this whole thing works at the top level," I told him. This of course delighted him. So, we used all the influence we could muster to get him out there, stay at the best hotel, and break all the rules to get him tickets for the screenings and meet the right people. This included the big producers like Arthur Cohen, the critics including Roger Ebert in the fore-front, and the top filmmakers of the day. We threw a gigantic Chicago party. Ellis brought his wife, whom he invariably and touchingly referred to as "my lovely wife Gillian."

Goodman saw what I was aiming at for Chicago, and incidentally had a great time. So next I said, "Now we should take you to Hollywood to meet the studio heads and the greatest people and see how we have to maneuver to get the films." Lovely wife Gillian came too. And in return he told me, "I can make your dreams come true." There were plans to build our own theatre, and Goodman was very pissed off when negotiations to buy a building to do it fell through. I felt he liked what I was doing.

Then came the summer of 1994.

I was invited to be on a film jury in Jerusalem, so Ellis said, "Yes, you go off to do your jury thing in Jerusalem, Michael. Everyone's doing a beautiful job." So off I went.

I came back to find the following headline in the Chicago business paper Crain's Chicago Business: "KUTZA TO BE CANNED." (Get it? Canned, like the Cannes Film Festival.) It was written by Lewis Lazare, who at that time was inclined to see himself as a kind of smarmy local Rex Reed. Nobody at the office seemed to know anything about it and everyone was very chatty except for our fundraiser Eileen Murphy, who one day murmured, her beautiful face stony, "Watch your back, Michael!"

That froze my blood.

So, I had to learn it from the newspaper. The plan was that I would disappear and be replaced by Marc Evans, my programming director. Marc was a gifted and charming young man, then in his twenties. Ellis did not really get to meet the staff, so obviously in this instance he had taken advice from Betsy, who had orchestrated the reorganization of the staff, which then moved to Ellis's hands-on control.

The festival treasurer was the distinguished lawyer Phil Azar at the law firm Arnstein & Lehr whose specialty was taxes and money matters. He had been concerned because of Ellis's dominance of the finances. When Phil complained, "I can't really do anything, because this man is totally controlling everything in the books," I could only reply, "Well it is his money, remember. And he is giving us major salaries, which we've never had before. So, we sort of like Ellis, don't we?"

Not only that, but Ellis had cleared out the longstanding debt, at a cost of $500,000. Ellis asked Azar to leave the board.

Now the papers began reporting that Kutza must go, because all the sponsors were supposedly leaving because of me. This was the first I had heard about sponsors not liking me. And now I found that some of my staff members were suddenly very much against me. It was incredibly hurtful.

Finally, there was a front page headline in the *Chicago Sun-Times* saying, "Kutza's Being Removed By The Board of His Own Organization." Whereupon I had a call from a lawyer friend of mine, Mickey Gaynor of the powerful firm Schwartz, Cooper, Kolb & Gaynor, a wonderfully strong and supportive guy.

"You need representation," Mickey said.

"Well, Burt Kanter just called," I replied.

"Yes, we're going to help you, the two of us, and we're going to remove Ellis Goodman for doing what he's doing to you."

Then it all just poured out of me.

"You know what, you guys? I'm reading the paper, and I'm finding I'm some kind of a bad guy. And he wants to take over this whole thing. I don't want any part of it anymore. This is a great hurt to me. I don't even know the details, and I'm really hurt, and I don't ever want to be involved with this organization."

"Are you crazy? It's yours, you founded it," Mickey pointed out. "You haven't done anything wrong. And this man wants it."

"Yeah," I said, "but I'm disgraced by whatever this is, whatever's going on."

Mickey was short and to the point: "I'm going to represent you."

Now the cost of a lawyer's time at this level is 400, maybe 500 bucks an hour. The next thing I know, we're listening to extensive memos from lawyers for Ellis at Katten, Muchin & Zavis, with endless conference calls with Ellis.

I found it devastatingly sad. I was hearing a pack of pointless lies about the reasons why I should be removed. And then when we turned to memos with my lawyers, Ellis would actually be there saying, "I have never said these things." He would deny everything for the record. Then we would go back to his lawyers and he would admit all the same things he'd just denied.

It was very confusing to me, because I knew nothing about law. But I guess this was just how it went.

Finally, Burt said, "Well, we will need to have a board meeting to decide if you're leaving or not. We'll need to decide if you'll be getting a pension or a lump sum payment. You've given all of your time and money to the festival. They should *not* let you leave empty-handed."

Mickey added, "Well, even if you are willing to leave, it's only fair that some provision should be made. So, there isn't anything happening, except that the board will vote on whether or not to keep you around."

I must have seemed a very ungrateful client.

"I don't ever want to go back!" I fumed to my lawyers. "I don't give a damn anymore!"

And then Mickey Gaynor said, "Why don't they just pay you off? Why don't you agree? Let's have them give you $250,000 to just go away. But Ellis doesn't want to do that. He wants you to leave in disgrace and then pay you nothing."

"I don't care!" I replied in disgust. "I don't want to ever hear the words Chicago International Film Festival again!"

"That's ridiculous," Mickey insisted. "You built it. It's very successful. Ellis wants it and he doesn't care if he destroys you in the process."

I think I lost about ten pounds that year and my hair turned gray. Mickey reviewed the relevant corporate documents and advised me that it was his opinion that neither Ellis, as chairman of the board, nor the executive committee he had set up had the legal right to fire me. That required a vote of the full board, at which a quorum was present. Mickey conveyed his opinion to Ellis's lawyers, who conceded. A board meeting was scheduled.

Finally, the day of the big meeting arrived. Before it, we met in Burt's office. He said, "Let's make a chart of the people who are going to vote for you and those who are going to vote against you."

"I have no clue," I responded miserably. "I assume they all hate me, even though I've done nothing wrong." I was being petulant and immature. Burt ignored me, responding, "Let's figure this out."

So, we added it all up and Burt concluded, "I think you have enough people that will vote for you." I was incorrigible, noting, "I don't care." But Burt was unmoved. "Yes, you're going to win this thing, because Ellis wants the power, and he doesn't care if it destroys the organization. But he won't succeed. Let's get down to the board meeting."

The meeting was held in the office of one of the board members who didn't like me. There was the board, which now included Ellis's lovely wife

Gillian, whom he had recruited. And there was Betsy, who had remained very close to Ellis.

As soon as the meeting was set to begin, Ellis insisted that some non-board members should leave. These included Gordon, of the restaurant Gordon Sinclair, and Phil Azar, even though as treasurer Phil might have been assumed to be legally required to be present. I think Ellis even tried to kick out those who had a legal right to be there. He demanded that Mickey leave, but it was pointed out that Ellis had his lawyer present, so mine should be permitted to stay, too. Bullshit intimidation tactics. The board voted to allow Mickey to stay.

I took my turn before the meeting began to say my piece and pointed out, "Well, Betsy is the manager of this thing, not a board member, and she shouldn't be in this room, either." I knew everyone loved Betsy, because she controlled them. Ellis was shocked and outraged, insisting, "But it's Betsy."

"Betsy isn't a board member," I replied coolly. "She cannot be here." Whereupon Betsy burst into tears and fled the room. But she had hardly closed the door when she returned and laid her resignation letter on the table. I found this weird, but it was becoming clear that all these things were planned in advance.

Enigmatically, the "lovely wife" Gillian burst into tears next. That was evidently *not* part of the plan.

At this point, I thought, "This is too much for me." With such high drama in performance, who is going to care about me? However, Burt Kantor was undeniably on the board and, as a strong, dynamic, yet quiet and subtle man, he provided the power to orchestrate this meeting.

With that, the meeting began. As chairman, Ellis announced, "Well, it's clear that Michael has to be removed blah, blah, blah." Burt quietly responded, "We need to take a vote and decide how many people here want Michael to remain…or you to remain, for that matter."

This left Ellis agitated, because he was anticipating a unanimous agreement that I had to go. But Burt patiently explained, "No, this is a board. An executive committee that is all here." Preparations for the vote began, but meanwhile a known Ellis supporter, Maryann Rose, declared, quite hysterically, "We have got to get rid of Michael because if we don't, he will start saying bad things about us. Because that's what he does."

I was thinking, "Well, of course she's right."

Then another man who for years I had regarded as a dear supporter, Albert Jay Rosenthal, a delightful little white-haired man, chipped in gently, "Yes, Michael must go."

My God, I thought, these are my friends, people I've worked with for years. With fickle friends like these, who needs enemies?

Nevertheless, the vote went forward. I think both parties were equally surprised when I had enough votes in my favor to survive the assassination attempt, leaving those who were defeated and outraged to flounce angrily out of the room and quit on the spot. Things were getting very interesting.

Ellis was flustered. Clearly, he was not accustomed to losing the game.

"Well," he said, "with this I will resign."

"No," responded Burt. "You'll have to stay on for the rest of your term, no matter how you feel about what's happened here. You will complete this year and we will have an interim committee to decide how to run this organization and fix what's happened."

I always wondered if Ellis respected or feared Burt. But at that moment, I honestly didn't care. He'd lost. I'd won. That was all that mattered at that moment.

The irony of this entire little farce was that the argument for getting rid of me surrounded sponsors walking out. It simply wasn't true. Nothing like that had happened, but I was so gullible that I didn't bother to check. Subsequently, when I spoke to American Airlines, for example, they said, "Well of course, Michael, if you're gone we're not going to stay there."

"But they said…" I replied.

And they said, "Well none of that was true. That was simply trumped up."

Marc Evans, my trusted programmer, was out of the loop surrounding all this. He had been sent to California to wait until it was all done, I was dumped, and he could come back and run the festival for the foreseeable future. Maybe this was a good idea. He would have run the festival well, and I had to leave sometime. But that doesn't alter the fact of the way it was done. To set out to destroy someone in this underhanded manner was neither right nor fair.

Anyway, I won the battle, but I still didn't want to go back in and be a part of it anymore.

"I can't, I can't face it," I admitted. "They had set out to ruin me, just for the pleasure of acquiring power, with nothing to lose."

But in time, I returned, very content to be reunited with the festival I'd created and its new staff and newest chairman Dan Coffey, a bright, young architect who restored old classic movie theaters and turned them into theatrical venues. What could be more appropriate for the film festival? We also invited Judy Gaynor (a.k.a. "Mother Courage") to return and save the festival as she had done before.

It's a lesson for all nonprofit organizations. I have watched chairmen destroy arts groups before. They have nothing but money, and they love the power it gives them. I've watched the same stupid things happen with dance groups, theatre groups, opera groups. And such coups d'état can sometimes bring very positive results, albeit with causalities along the way. American Ballet Theatre was created by plots that pushed the great Russian dancer and ballet master Mikhail Mordkin out of the company he had established, and which till then bore his name.

And there were positive results for others in the attempted Chicago coup. Evans, who was in California waiting for me to go, did not return, because I had been brought back. And he started his career over again. He was an intelligent and sincere young man and worked hard to find his way to the top. He became president of the Motion Picture Group at Paramount Pictures, and is happily married with two children. We sometimes get together. Life is too short to perpetuate resentments.

I often see Ellis Goodman and his lovely wife Gillian at the theatre, and I hug them warmly. which I'm sure freaks them out. Ellis continues to involve himself enthusiastically in the arts, while continuing to expand his business interests, recently into real estate. He has produced films and major stage shows and has written three novels.

Years later, the great Charles Benton was my festival chairman. I respected him a lot, but I felt somehow offended one day when he said to me, "The best thing that could have happened to you was that whole episode. You've become a better person because of it."

I was insulted by that concept at the time, but later I thought about it, and today I completely agree. If not a better person, at least this made me a different one. Now, I don't risk trusting anyone. But, yes, it's made me stronger, and I certainly value friendship and how far I will go with a person more than I had.

I look for motives more than most people now. And, though I hate to confess it, when I trained my next staff, without directly telling them, I always guided them to look for things that might ultimately hurt them. I don't want other people needlessly to go through what I experienced with this unpleasant story.

Years later at the 50th anniversary of the festival, I received a giant, expensive bottle of champagne and this note: "Michael: Good luck in this year's festival. Here's to the stunning accomplishment of 50 years and a deep thanks for everything you did for me and prepared me for. Best, Marc Evans."

Yes, miracles happen.

The German filmmaker Wim Wenders, a major figure in New German Cinema. (Robert Dowey)

Oscar®-winning producer Tom Rosenberg (*Million Dollar Baby*) at our festival in 2012. (Robert Dowey)

(left to right) Me with novelist and socialite Sugar Rautbord and auction executive Leslie Hindman. (R.L. Restaurant Staff)

It was pure joy when Tom Cruise (left) surprised Steven Spielberg as we honored the Oscar®-winning director in 2006. (Robert Dowey)

An admirer examines Oscar winner Sophia Loren's signature (and possibly more) when we honored Sophia at our Chicago Festival in 1990. (Matt Gilson)

The Italian film critic Guglielmo Biraghi appears somewhat distracted by something he sees on guest of honor Sophia at our festival. (Matt Gilson)

The Oscar®-winning acting great Sidney Poitier
at the Chicago Festival in 2008. (Robert Dowey)

Always the clown, Will Ferrell takes a selfie of
those capturing him on camera. (Robert Dowey)

Two-time Oscar® winner Ron Howard beams at
our festival in 2010. (Robert Dowey)

Director Robert Altman enjoying his time at the
Chicago Fest. (Robert Dowey)

Academy Award® winner Nicolas Cage walks the carpet
at the Chicago Festival in 2005. (Robert Dowey)

Actor Malcolm McDowell of *A Clockwork Orange* fame. (Robert Dowey)

(left to right) Larry (now Lilly) Wachowski and sibling Andy (now Lana) Wachowski of *The Matrix* fame. (Timothy M. Schmidt)

Oscar® winner Helen Hunt strikes an ebullient pose. (Timothy M. Schmidt)

Catching up with the esteemed Italian film critic
Gian Luigi Rondi. (Michael Kutza Archives)

Three-time Oscar®-nominated actor Edward Norton
at the Chicago Festival. (Timothy M. Schmidt)

Two-time Oscar®-winning actor Dustin Hoffman (right) and
me in what seems to be a battle of the noses. (Ian Sklarsky)

The film critic, festival director, author and dear friend
David Robinson. (Timothy M. Schmidt)

It's opening night of our 49th Chicago International Film Festival in 2013, and a sellout crowd is already milling about – always a good sign. (Robert Dowey)

Chaz Ebert, Roger Ebert's dynamic and lovely wife. (Timothy M. Schmidt)

Me with (left to right) columnist/blogger/media personality and the toast of the town Candace Jordan and Randy Crumpton, lawyer, board member, and good friend of the Chicago Festival. (Robert Dowey)

The versatile actor Taye Diggs makes a point at our festival. (Robert Dowey)

The Scottish actor Alan Cumming. (Robert Dowey)

(left to right) Me, Jackie Rabin and her Chicago Festival board member husband Ed, and co-owner Joey Mondelli at his incomparable Italian restaurant La Scarola. (La Scarola Staff)

Oscar®-nominated director Norman Jewison, who makes comic fodder of the fact that despite his name, he's a gentile. (Robert Dowey)

The mega-talented performance artists Laurie Anderson (left) and Ken Nordine spending time at our festival. (Timothy M. Schmidt)

Polish film director Krzysztof Zanussi. (Robert Dowey)

The cult Italian horror film director, producer, screenwriter,
actor and critic Dario Argento. (Timothy M. Schmidt)

Michael Rooker, who starred in *Henry: Portrait of a Serial Killer* in 1986, remains a cult favorite decades later, as he discovered from all those who lined up at our festival to pose with him for selfies. (Timothy M. Schmidt)

Actor Dudley Moore (left) and director Blake Edwards having a good old time at the Chicago Festival. (Robert Dowey)

Academy Award® winner Geoffrey Rush. (Timothy M. Schmidt)

Two-time Oscar® winner Anthony Hopkins (right) and
his actress-producer wife Stella Arroyave. (Robert Dowey)

Hanging with the towering British actor
Christopher Lee (right). (Robert Dowey)

It was clear from the beginning that Anna Nicole Smith and
model Mark Kleckner had a smoldering chemistry after meeting
for a photo shoot. It was obvious to my own camera, too. They hardly
needed to pretend to be attracted to each other. It's all over their
faces and body language. (Michael Kutza Archives)

13 – SINGIN' IN THE RAIN

I know you've probably never heard of Albert Johnson, but in the Seventies, he was Mister Film Festival. For many years, he was the artistic director of the San Francisco Film Festival and a world traveler to just about every festival there was. It got to the point that many of us suspected he was a secret agent for the FBI or the CIA.

There was certainly something mysterious about Albert. It seemed that whenever he was at some exotic film event, that country was just about to head into a civil war or other major conflict. It was the perfect cover, a film festival director, sorta like Josephine Baker when she was recruited as a spy in France.

Oh, did I forget to mention that Albert was also African American like Josephine? Black and born in Harlem, New York. His father was a doctor. His mother was once a dancer at the famed black nightclub The Cotton Club. Albert was born a song-and-dance guy at heart, and he knew everything there was to know about Hollywood musicals and old-time Hollywood stars. He put the San Francisco Festival on the map with his on stage star events.

At the San Francisco Festival afternoon performances, he would have on stage Fred Astaire, Judy Garland, Bette Davis, Howard Hawks, and Busby Berkeley. He put the festival on all of our radar screens with his uncanny ability to attract A-list talent. In fact, he became a little too popular for comfort for the festival's actual head programmers, Peter Scarlet and the figurehead executive director Claude Jarman, Jr., a onetime child actor in *The Yearling* (1946) who was a major donor to the festival.

Albert wanted more free programs for the students and was ultimately forced out, grumbling about how "commercial interests undermined artistic standards." At least his magic continued as a teacher at U.C. Berkeley.

I met Albert at the Berlin Film Festival. That festival's hospitality department offered guests, as part of its program, a little tour to see the famous, or sometimes infamous, nightlife of Berlin. The first stop was the famed KitKatKlub, the

140

telephone bar we all know from the movie version of *Cabaret*. The place had changed at lot since the old days. You could come and party and even leave your clothes at the door! I'm not joking. There were many festival directors at this particular party.

Albert introduced himself as "San Francisco" and me as "Chicago." I'm telling you, the guy was quite the character. After a while, Albert suggested, "Let's try the next club," as this gathering was really getting strange.

Next stop on our degradation journey was a place called the WuWu Bar. This whole adventure was taking place during Mardi Gras time in February, so many people were in all sorts of fancy dress. You also had the French contingent on the nightclub tour, directors and stars dancing and having a glorious time.

I suddenly found myself on the dance floor with my old Chicago Festival pal Jeanne Moreau. French actor Jean-Claude Brialy was dancing next to us. He was cute but very drunk, and he cut in and said to me very matter of factly, "I would like to fuck you."

"I don't think so," I replied, "and it's time to get out of this place." So Albert said, "Let's go to the next stop on this tour." We were done with the WuWu. That brought us to the KC Bar, or Kleist-Kasino. It was famous because it's where Nazi soldiers hung out back in the Forties, and it was just across the street. It turned out to be a very interesting place, a hustler bar where you could just pick up people and pay the fee. No muss, no fuss.

Albert fancied this sailor boy, probably in his twenties, and he told me, "Michael, we're going to take him back to the hotel."

"Albert, I don't think so," I reasoned. "It's 3 in the morning, it's snowing, and this kid is so drunk he just peed all over his nice white sailor suit."

But Albert was undaunted.

"Let's get a taxi and go back to the Hilton," he suggested, which struck me as a terrible idea. But that's just what we did.

I was so paranoid and shy as I walked in out of the snow with a black man and a 22-year-old sailor covered in pee, strolling into the lobby of this chic hotel in the middle of the night and asking for my room key. I was so nervous, but there was no problem. We just got straight into the elevator and went to my room.

"Throw him in the shower," Albert ordered.

"Throw him in the shower with his clothes on?" I responded, incredulous.

"Yes, you gotta wash the clothes off," Albert said calmly, as if he did this every day. For all I knew, he did.

"This is crazy!" I pointed out.

So, we did throw the drunken, pea-soaked sailor in the shower. I thought, this is insane. I have to get in the shower with him to take his clothes off, because they're so wet. Needless to say, we had a compromising situation here, having to take the clothes off this kid in the middle of the night in a Berlin hotel shower.

Mind you, I had already taken my own clothes off, as I was not going to get them soaking wet.

I decided to point out the obvious.

"You know, Albert, I don't think this is my scene, and I'm leaving you two here. I'll go to your room down the hall." Which I did, leaving Albert to have his way with the kid in the shower.

The really fun part of the story came the next morning, when I started receiving telephone calls and answering them. I was in Albert's room, so anyone who called now assumed that I was involved with Albert, while Albert was accepting phone calls in my room, and anyone calling me also assumed that Albert and I were a thing. It somehow never occurred to me to tell him not to answer my phone, or for me not to answer his.

This went on for a few days, but it did allow me to bond with some curious people on the phone. I had never met the director of the New York Film Festival, a rude, gossipy man named Richard Roud. He also ran the London Film Festival. I also met on the phone one David Robinson, the film critic of the Financial Times, and John Russell Taylor, film critic for The Times of London. Both had to be thinking they were talking to Albert's new lover: moi.

It took a few days to explain the situation, and I ultimately became best pals with Albert. I invited him several times to the Chicago Festival to be a guest interviewer. Whenever Albert screened a musical film, he would also sing and dance a number from the film on stage. He was terrific and terrifically entertaining.

Sadly, Albert also became our first Chicago Festival fatality in 1998.

I saw him that spring at the Cannes Film Festival and said, "Albert, we're doing a tribute to Pam Grier this year, and I'd like you to interview her, since you're an old friend." He agreed, but I noticed that he wasn't well, moving much more slowly than I remembered, seeing fewer films at the festival, and just not being Albert. He was trying to hide it.

So along came our festival in October, and we had the Pam Grier event. Albert phoned right from the airport that he'd arrived and was going directly to the hotel. That was the last time we spoke. I would leave phone messages at the hotel and they'd say, "Oh yes, he was here in the morning and went to visit the hospital but returned to his room." Still no word from Albert.

The show must go on, so we found a songwriter/film promotion guy in the audience who knew Pam and her work. He agreed to go on stage and do the film clips and interview. Still no word from Albert. I asked the hotel to please go check on him. It seems he'd died in his room after returning from the hospital. I was in shock and desperately wanted to see him, but the police said only next of kin. He was taken to the morgue.

It turned out that Albert was dying of prostate cancer. He knew it, but kept it to himself. He insisted he was coming to Chicago and doing the event for Pam, no matter how much pain he was in. The obits had him passing from a heart attack. Not true.

We all loved Albert. He was a joy to be with, even if he was Josephine Baker and an undercover spy all rolled into one. It was a combination he would have been pleased to embody.

14 – IT'S A MAD, MAD, MAD, MAD WORLD

So many figures have entered my life and had an incredible influence on me. Keep in mind that I was still just a kid from the West Side of Chicago who was now rubbing shoulders with legends like Federico Fellini and Charlie Chaplin. I was working at the Venice Film Festival the summer Chaplin was honored. He was receiving a Lifetime Achievement Award in a rare public appearance and wasn't in great shape, but able to get around.

I introduced myself and presented my Chicago Festival business card, which inspired the legend to ask, "Is that my image on your card?"

"No," I replied, suddenly fearing a likeness lawsuit, "it's Theda Bara."

"Pity," Chaplin observed.

He went on stage, cried, and was utterly charming. I also received a Silver Lion honor that year for introducing American independent cinema to the Venice Festival, on the same stage as Charlie.

Many years later, Guglielmo Biraghi became a major mentor in my life. He was founder and director of the Taormina Film Fest in Sicily. I was on the feature film jury, and we became lifelong friends. In the Eighties, he was named director of the Venice Festival. He was a fascinating man in love with life, films, seashells and exotic actors and actresses. I had a chance to meet most of them.

Guglielmo introduced me to Lina Wertmuller and Sophia Loren. He was on our various film juries for years in Chicago. It was there that he momentarily fell for a German actress named Lisa Kreuzer, who was once married to Wim Wenders.

One of the really funny events in my life happened when Guglielmo was directing the Venice Festival and he couldn't get rid of Lisa, so he invited her to Venice. Big mistake. Lisa was in love, but Guglielmo still had his Italian actress wife (Annabella Incontrera), whom he never could get rid of even after divorce. She was always roaming around needing money, even after many more failed marriages.

This story unfolds like a Blake Edwards movie.

My dear friend Patrick, a film critic and festival director in Belgium, was in Venice at the festival. He'd just broken up with a lover that very night and was extremely drunk trying to forget about him. We were all at the Excelsior Hotel, the festival's official spot. Patrick was alone and as described, inebriated. He was being plied with still more drinks by the director of the Berlin Film Festival, to mitigate his sorrow.

The Berlin director had only one thing in mind: having sex with Patrick. To save Patrick from Mr. Berlin, Guglielmo, seeing the situation unfold, took Patrick up to the safety of his own hotel room. Patrick straightaway passed out on the bed.

The next day, Lisa arrived to find Patrick in Guglielmo's bed and stormed out, clearly not understanding the situation. Then came more drama with Annabella drunk and banging on the door, demanding to enter. Announcing to the maid that she was Guglielmo's wife, the housekeeper let her in. She found Guglielmo and Patrick together in the same bed and naturally went predictably ballistic.

This made Guglielmo officially G.A.Y. in everyone's mind, of course. Not that there was anything wrong with that.

How did it all ultimately play out? Well, Patrick safely returned to Belgium and married a sweet, talented architect, living happily ever after. Annabella remarried. Lisa returned to acting in Berlin, alone. Guglielmo lived to run many more festivals and continued to write books on seashells and films. And he remained happily straight.

Lesson learned: no good deed goes unpunished, predictable when it involves hotel beds and alcohol.

--

In 2001, I dedicated the Chicago International Film Festival to Guglielmo, who had just died of cancer in Rome. The man was omnipresent at our festival as the president of our feature film jury and as coordinator of our relations with Italy and the Italian film industry for so many years. When you speak of film knowledge and cinematic personalities, his name comes up everywhere. In some way, in fact, he touched nearly every international film festival and community around the world for more than 30 years.

Guglielmo had also served as the staff film critic for the Rome-based daily paper *Il Messaggero* since 1953. In addition to all that, he wrote and directed theatrical productions. What he gave me in Chicago was simply immeasurable. He brought a calm with him that was contagious, even amid the barely controlled chaos of an international event like mine.

You could have temperamental film directors or jury members ranting and raving in six different languages, about six different problems – in fact, Guglielmo spoke five languages – but nerves were always calmed and crises resolved when he intervened. The man was such a diplomat that he could have been the president of the United Nations if he'd so chosen.

In addition to that, his knowledge of film was vast and invaluable. And it didn't hurt that he knew innumerable great filmmakers personally. My film heroes were his friends: Fellini, Bergman, Visconti, Welles, Ford, Minnelli, Truffaut, and on and on. Yes, I got to meet many along the way as well. But Guglielmo was actual friends with them.

Guglielmo repeatedly called me from Rome when he was very ill, having never quit smoking, which would ultimately be the death of him. To the last, he shared endless ideas on how to make things better or smoother or more international for me in Chicago. And he always ended his call by saying, "Best wishes from your younger brother in Italy."

He loved Chicago and he touched us all in so many ways. We rejoiced in his presence, and to this day, I miss Guglielmo so much. I'm crying all over again just writing these words.

If you're lucky, you have one friend like him in your lifetime.

I was lucky.

--

I invited George Cukor to Venice in 1982 to receive the festival's highest honor, the Gold Lion, and to unveil as a world premiere the newly restored three-hour version of *A Star Is Born*, George's 1954 masterpiece. He was very frail but agreed to travel and even to do a press conference after the film.

Everything went well as far as anyone could tell, but George was losing it mentally and physically that year. We were walking down this long hallway after the press conference, just the two of us, when suddenly his trousers fell to his ankles. He had been losing so much weight at that time that the belt just didn't keep his pants up. He was skin and bones under his suit jacket. It was very funny, but not.

I managed to get George to the hotel lobby and left him standing there while I got his hotel room key from the concierge. He wandered to the giant fountain in the lobby, heard the water flowing, and decided that he needed to pee. So, he did it right there in front of everybody. He was oblivious to it all and just kept failing from that point on. He later was fine receiving his award on stage. He flew back to L.A. the next day. It was the following year that George died of a heart attack, at 83.

--

The first Manila International Film Festival in Philippines came along in 1981, and I was asked to join on as an advisor. This was a real adventure as I had never been to the Philippines and didn't know what to expect. Imelda Marcos, First Lady of the country for 20 years and with a 3,000-pair shoe collection, wanted a film festival, and she produced a gigantic earth-shattering event.

Mrs. Marcos had built a temple to the cinema on an elevation across from the famous five-star Manila Hotel. Her film building was designed to look like the Parthenon, costing $25 million but ending up a mausoleum in many ways. It had nothing but construction problems along the way.

One major disaster came when a superstructure collapsed either from the rush of building it so quickly or an earthquake that killed a few hundred people inside wet concrete. Imelda would not stop the building as it had to be completed for the opening of the festival. Workers would not return, as many of their family members were now buried in the place and it was therefore haunted with their spirts. In her Catholic brilliance, Mrs. Marcos convinced her favorite cardinal to come and exorcise the place of bad spirts and ghosts. His name was actually Cardinal Sin, I swear.

The cardinal exorcised the place and work continued despite the death toll. It all finished on time and the festival was about to open but there were some unfortunately new real dangers. Imelda and her husband Ferdinand faced civil unrest and rebellion as a result of many issues facing the citizenry.

Manila was under martial law, and bombs and bomb threats erupted all over the city. The possibility of an explosion was quite real on opening night of the festival. Mrs. Marcos had flown in some major movie stars, and the opening night film was the multiple Oscar winner *Gandhi*.

Imelda was a genius. She ordered silent fireworks for the opening. No bangs, just fabulous light displays in the sky. I was in the Manila Hotel just across the way from the film center structure and chose to watch the entire opening on TV, just in case violence erupted. Death was not in my contract, and I was really scared. It all went off without a hitch, however.

On a lighter note, one of the guests at the Manila Festival was Motion Picture Association of America President Jack Valenti. He was such a big shot that he actually was staying at the Marcos Royal Palace, which Jack proudly told everyone. It didn't last long.

Jack had a great head of hair, perfectly combed and sprayed and blown into place. One night, he left his hairdryer on by accident, went to a gala party, and started a fire in the palace. This was not appreciated. He was removed to a hotel immediately.

Since I had never met Mrs. Marcos before, I was always impressed with the lady. At the first gala dinner, she rose and out of the blue started singing a song to an important guest at the main table. It was quite something, but when suddenly the orchestra was playing and a choir chimed in, I realized I'd been had. She had the whole thing staged.

I went to a few more galas and saw it again and again. I realized she was a professional singer. And very good by the way. I knew she was a dictator's wife and maybe even a dictator herself and not a good human being, but I have to admit I was nonetheless charmed by the woman and liked her very much. She also liked me and had me do a few on-stage interviews with film directors and producers. One was with a producer named Arthur Cohn. I told her I had no idea who he was.

"You *will* know by tomorrow," Mrs. Marcos demanded.

From that day on, I became lifelong friends with Arthur and even worked together on two of his films years later. He's a documentary producer and six-time Oscar® winner whose big break had come in producing the 1970 film *The Garden of the Finzi-Continis*, directed by Vittorio De Sica.

I would ask Arthur for suggestions on upcoming films that he had seen, and he would ask me for advice surrounding his latest films.

In 1997, Arthur came upon a new young director, Ken Selden, at the Sundance Film Festival and developed a film entitled *White Lies* that dealt with interracial relationships. He had a problem with the musical score and asked me out to Hollywood to advise on new music approaches. While doing that, I suggested they should shoot some additional scenes to make the film a bit sexier.

The *White Lies* theme was already pretty adventuresome for Arthur. You had a young Caucasian woman (Julie Warner) having an affair with a young African American museum worker (Lawrence Gilliard Jr.). Lots of stars, but it was too tame for me. I also suggested adding some more local color to the opening to establish Harlem and New York City.

Arthur told me to come to New York. I rented a video camera and shot some B-roll to use under the opening credits and also designed the opening titles. He was in great spirts and fun to work with. I was at the Swissotel next door to where he was staying, at the Regency Hotel on Park Avenue.

One day while we were together, something a bit alarming happened. Arthur suddenly wouldn't speak to me, avoiding me in the hotel lobby and acting very rude. It was just incredibly weird. What could I possibly have done to offend him? I felt like, well, OK, I guess I've just been fired. I'd seen Arthur's moods before, but this was different.

The next day, all was well again at breakfast. I got around to asking him what was up. He laughed and explained, "It was the Sabbath. I was on my way to temple. I can't speak, open doors, or deal with people." Of course! Arthur, the son of a rabbi and a devout Orthodox Jew, followed the Friday rules to the letter.

This reminds me of the story of a very young Frank Sinatra actually working part-time at the Regency Hotel to open doors and push elevator buttons for the Orthodox Jews on Friday. The slang term for it is "Shabbos Goy."

I worked on Arthur's latest film in 2018, *The Etruscan Smile*. He needed director and a star. I wanted Mark Romanek (*Never Let Me Go* and *One Hour Photo*) to direct. He was a Chicago boy and I liked his work, as did Arthur. Alas, Mark was making more money doing music videos and TV commercials and wasn't interested in making another feature at the time.

I also wanted Sean Connery for the lead. He was perfect and perfectly expensive, and wouldn't travel to Europe to shoot the first half of the movie. Arthur zeroed in instead on Brian Cox. Turned out to be an excellent choice.

Arthur has boundless energy as a producer and is constantly looking the next great script to produce. As I write these words, he's also 95. I hope he never dies.

--

Getting back to Manila, it was in the Philippines that I experienced my first earthquake. Walking through the lobby of the Manila Hotel, I felt suddenly uneasy, almost sick and dizzy for a moment. Then I saw parts of the ceiling molding starting to crack and fall and the giant chandeliers starting to sway. I thought it was a bomb. This, however, was silent and terrifying.

Then just as suddenly, it stopped.

Being Catholic, I wanted to go to church, especially a famous church where the priest was curing someone and tearing the sickness out of their body. I got to see it firsthand. Yes, that's correct. I couldn't believe what I was seeing. Blood was all over the place, and something was being ripped out of the person's body, the cancer or the whatever. I bought into it and will never forget it.

Of course, there were also the nightclubs of Manila. The sex shows with the added feature of you choosing the dancer and having sex with he or she upstairs at the establishment. I believe it's called prostitution. Most of the dancers were of age, whatever that means in Manila, and they lived in the place, like a dorm. This is how the kids made a living in this very poor Philippine city. Was it sexual slavery? I couldn't say.

Meanwhile, you had the outrageous grandeur of the festival just down the road and wrenching poverty everywhere else. No wonder civil unrest was running rampant.

By the way, in case you were wondering, renting a human cost $25.

15 – JURASSIC PARK

In 2000, I had the honor of hosting the Academy Award®-winning British director Richard Attenborough. He had directed some terrific films, including *Young Winston, Cry Freedom, A Bridge Too Far,* and of course *Gandhi.* But in America, we had no inkling of his earlier career as an actor in Great Britain, which had seen runaway success on stage and (debuting in 1942, at 19) on screen.

Attenborough and his wife Sheila Sim were the original stars of the world's longest-running play, *The Mousetrap,* which played nonstop from 1952 through COVID-19 in 2020. He appeared in almost eighty films, in Britain and America, notably including 1966's *The Sand Pebbles.* He became a producer and, in 1969, a director with the inventive musical *Oh! What a Lovely War.*

But Attenborough had his mistakes, too. He took on Broadway with a screen adaptation of the hallowed *A Chorus Line* in 1985. Big mistake. The 1992 film *Chaplin* should have been great, but it wasn't. He just got so carried away with adding endless starry bit parts, including Charlie's daughter Geraldine playing her own grandmother, Hannah. Richard added more substories into the script when all he needed was the genius of Robert Downey Jr. to carry the entire film.

Apart from managing a busy career, Dickie was a major public figure in his own country, and eventually appointed a Baron, so that as Lord Attenborough he had a seat in the Parliament. He was also a tireless worker for charities and the chairman or president of around fifty organizations. He was always ready to interrupt filming to sort out one of their financial problems.

But the key to Attenborough's breakthrough in America was *Jurassic Park* in 1993. I know he didn't direct it, but it had suddenly given a quite new luster to his stardom on these shores.

Moreover, my dearest friend in Chicago, John Lanzendorf, had been working with a dinosaur project and writing a book on dinosaur art. Chicago's

Field Museum was about to unveil SUE – a giant, fully restored dinosaur. It seemed the proper opportunity to honor Dickie and bring him together with SUE. That way, our film festival audience would want to know about him and his films and acting career. It worked as planned.

It was all made possible by another of those mentors who have played an important role in my life, David Robinson. *Chaplin* was based on Chaplin's autobiography and David's still-standard biography, *Chaplin: His Life and Art,* and David was a fulltime adviser on the film. He convinced Richard to come to Chicago on a flight, bringing with him his regular collaborator Diana Hawkins. She was a protective and unsmiling lady who had worked as a film publicist on Dickie's early films but had advanced to production roles and was credited with the "story" on *Chaplin.*

At first, I assumed Ms. Hawkins was his wife, and subsequently asked no questions.

Richard was a dream when he hit town. We dragged him to the Field Museum to view the new SUE. Since our Chicago audiences were relatively unfamiliar with Attenborough's acting and directing, the major TV and press coverage celebrated him as that famous and rotund professor who welcomes you to Jurassic Park. Then Chicago's public suddenly got it. He played it so well for the TV cameras. He was a nonstop actor and a terrific sport.

So, Dickie and Ms. Hawkins were settled into their penthouse suite atop the famed Hilton Hotel & Tower in Chicago's Loop downtown. They only used one of the two bedrooms. I went up there to welcome them with David, who had flown in with them to comfort Dickie, who didn't like doing these trips and hated festivals. But David had made it work.

Then, it happened.

It was mid-afternoon, and the electricity went out in the entire downtown area surrounding the hotel. A power grid burned out, it seemed. It was amusing at first: it was afternoon and we had great views of Lake Michigan from the many windows of the penthouse. The elevator was not working, so we snacked on the giant welcome basket of fruit and cakes that the hotel had sent up to the suite. Richard's talk wasn't until later that evening, in a museum, so all promised to be fine.

Then darkness fell. The fancy hotel had no power, not even hall exit lights to find the stairs, and we were on the very top of this giant hotel. I think the staff stumbled in with some sort of emergency lighting. All they could come up with was glow sticks.

"Well, let's open the bottles of wine they sent," Dickie finally announced.

It became a rather amusing, if unnerving, tea party.

"It reminds me of the war," he said. All I could think of was how do I get back to the festival and see what was happening in the city and at our film showings...and what about the tributes that evening? I looked out to the hotel's halls and saw total blackness. Nothing.

It wasn't panic time yet. Robinson, Attenborough, and the lady were all having a fine time. But I was starting to panic. I said goodbye to the gang and started feeling my way down the pitch-black hallways looking for some sort of exit staircase. I heard the scattered voices of others doing the same thing, people walking down the stairs in complete darkness, feeling the walls and calmly descending 29 floors and out of that building to some faint light.

I finally got to ground level and took a taxi to the festival, which was all fine, since that part of town was not on this same electric power grid and the truth was the festival could run itself just fine without me lurking.

By 9 p.m., all power was restored, and we could calmly get Dickie and Ms. Hawkins to the museum. The audience was modest. The tribute was hosted by John Russell Taylor with film clips. There was a personal tribute from Robert Downey Jr, recorded from the prison cell where he was confined in a substance abuse facility. It had been quite a day.

Dickie and Ms. Hawkins flew home a few days later. In the interim, we presented Richard with a sitting with our great Chicago photographer Victor Skrebneski, who made some magnificent portraits of the great man. We shipped them off to Dickie in England and he received in return a curt message from the assistant, saying that they had been on the hook for a customs charge on the photographs and asked if the festival would kindly reimburse them for the cost. We did, though not without a raised eyebrow.

Later, Ms. Hawkins wrote an affectionate biography of Lord Attenborough.

16 – LADIES OF THE CHORUS

When it comes to actually running a giant enterprise like the annual Chicago International Film Festival, I, of course, made it seem like I did it all. The truth is that behind the scenes existed a remarkable team of devoted, motherly women who believed in me and the festival so much that they gave up their own family and personal lives to support it so fervently.

Yes, that's how charming I was back in the day. You must keep in mind that at the start, we were a mom-and-pop operation. My father put a very smart woman in place to watch over the finances, since it was his money and I was just a kid who was destined to become a medical doctor to follow in my mother's and father's footsteps once I got this film thing out of my system.

Oh yes, about that smart woman. Her name was Ramona Santelli. My father trusted her. I never questioned who she was or why or how she drove a Mercedes. She ruled the office and the tiny staff with a firm hand.

After a few years, it was crystal clear I wasn't going into medicine, so we did a search for a proper manager. Along came Lois Stransky, a serious, delightful little lady, to run the show. Professional in every sense, she wore a foam neck brace all the time because of some past automobile accident, I assumed. I would later find that she was having twice weekly secret sexual adventures mid-afternoon with the editor of a local newspaper.

I ran into said editor at an event, and he commented, "I like that Lois of yours. She fucks like a bunny!" Thank you for that information, I thought, and it did make me wonder about the foam neck brace in a new way. As long as she got the job done at the festival, her personal life was her own. She got all the local foreign consul generals involved with us to help with events and visiting dignitaries.

When Lois became seriously ill, we found another motherly successor to replace her: Judy Gaynor, a favorite of the then-chairman Charles Benton. He knew her because she managed various political campaigns and felt she could run the show perfectly as the festival was growing. He was correct. Her love was cooking, and entertaining. Having guests over to her giant home in Evanston was her joy. Jack Nicholson smoking a joint in her garden made her life complete.

Judy loved the traveling part, too. She came to the Soviet Union with me as my plus-one when I was on the jury at the Moscow Film Festival. She developed governing boards and held the hands of staff members and volunteers to get them through all the trials and tribulations of the festival. Visiting her first Cannes Film Festival was to be her big trip, but she unfortunately caught hepatitis just days before and was quickly replaced with Suzanne McCormick, our film programmer and grant writer, as the trip and Cannes were already in place.

It played out almost like *All About Eve* in every way. Suzanne had been hired by Judy as the film programmer, but after Cannes she was in charge of the whole festival. Older boardmen loved her. She spoke four languages and was a professor at the University of Chicago. Having once been married to Irving Thalberg Jr. wasn't so bad, either. (Who was Irving Thalberg Jr.? Why, he was the son of Irving Thalberg Sr., a pioneering producer in the early 1900s and a tireless boy wonder who packaged movies ingeniously and profitably long before it was a thing. Thalberg Sr. was also married to Oscar-winning actress Norma Shearer.)

Suzanne was a pro and so good as an organizer that when the Los Angeles film festival FILMEX lost its leaders to AIDS, she was recruited by its new leader (as she knew him rather intimately from the Cannes Festival) to run the show. I was hoping to go to Los Angeles as well, but he only wanted her.

It was tough losing her, but a young, dynamic filmmaker named Colleen Sullivan came along. She started by answering the phones and moved up, replacing another managing director to run the entire show. Colleen was there for some of the biggest events honoring stars in the festival's history: Sophia Loren, Jack Lemmon, Rod Steiger, Kathleen Turner, and Oliver Stone. She traveled the world, did the endless festival circuit, and selected the films alongside me.

Eventually, the stress of having no money and pressures from the Internal Revenue Service and creditors, and the ugly Ellis Goodman hostile takeover attempt, did Colleen in. She was offered an excellent post at a famed PR firm in Chicago, and she took it. She then married and moved to London. I recall that she hated me at the start, but I eventually won her over.

Who better to start the show all over again than our best-ever mother figure, Judy? Later came Sophia Wong Baccio, a lady with a background from Hong Kong and a mother who was a movie star back there. I'm not sure what drew her to the Chicago Festival, but she was serious and unsmiling but otherwise picture perfect in every way.

Poor Sophia was up against a chairman who was unpleasant and pushed her too far too often. She was strong but had a breaking point with him and with me, too. When it finally got to be too much and she resigned, I recall her saying, "I want to be the founder, not the executive director."

"Sophia, it doesn't work that way," I assured her. "You need to create something to be a founder."

She had various leadership jobs after that but none to her liking until she started her own little monthly pop-up Asian film festival where she could be The Founder. This was a big deal to her. She could now be her mother, that once famous Hong Kong movie star. It seemed to work well for her.

But the show must go on, and once again the chairman Dan Coffey placed a favorite young lady in the post: Dulcie Gilmore. Dulcie was a fascinating, clever operator. After leaving her role managing a famed downtown theatre in Chicago, she came on as our executive director, since our then-chairman wanted to build us a permanent home for the festival, and she was retained to pull it all together. That was the plan, anyway.

Unfortunately, she wasn't a hard worker and couldn't delegate a staff. We really couldn't afford an expensive, non-hardworking executive director who wanted a top salary. It was remarkable how she could charm some of the board members to buy into her cause. I found her a great disappointment. She was finally asked to leave when the board saw the light.

Betsy Morris was hand-picked by the next chairman, the scheming Ellis Goodman, to re-imagine the film festival without me. She was also clever and at times very spacy. I never questioned that part of her, or what she might have been under the influence of that made her that way. All I knew was that the chairman at that time adored her. I never questioned that, either. I just knew she wasn't what she pretended to be.

We all bought into Betsy, as she had increased our salaries to new, unheard-of heights with the chairman's money. They basically bought us all, and why question anything? The big takeover didn't happen. It all fell apart with the board voting out the chair instead of me. So all was good.

Who needs all this politicking? Egos and more egos. The nonprofit world is a killer when it comes to not-so-dedicated, supposed lovers of the arts.

Along with all of those memorable staff members came a bunch of volunteer women who ran the show from a different angle. They included promotion manager Mary Sweeney from WCFL radio. She was an insanely fun lady with clever promo ideas every day to keep the festival in the news back then.

Mary Ann Josh was a remarkable friend who created a company called Events Alive. We did fashion shows, galas in museums, you name it, all to raise money for the Chicago Festival. There was a spectacular outdoor fashion show in the First National Bank Plaza around a giant fountain shooting water sky high. The party was called Million Dollar Baby long before Clint Eastwood used that title.

That party had over $1 million worth of diamonds and furs. I recall the guests all got rather drunk outside around the fountain and one guest even jumped in the fountain. It might have been Mary Ann, who was just the most positive person I had ever met. Even dying of cancer, she never lost her sense of optimism, determined to beat it to the bitter end.

It was Mary Ann who introduced me to a young African American woman from the Compton Advertising Agency. Her name was Dori Wilson. She started out as a fashion model and later became a master at PR. Just a kid when we met, she and I "dated" for years. I was the perfect "walker" for Dori while she was searching for Mister Right, who of course had to be white and blond.

I introduced Dori to our photographer Skrebneski and he made her a big deal in his ad campaigns for Virginia Slims cigarettes and department store ads he was doing at the time. She moved up and up at agencies. In those days, she was filling a quota for diversity.

Dori finally decided to start her own PR firm but never worked much with black talent or clients. Only later in life did she come to realize that she was missing out. She did fall in love with some cool guys, all white of course. A one-of-a-kind friend, but not an easy one, Dori was simply fabulous. She left us all suddenly early in 2021, never saying goodbye.

Donna LaPietra was a wizard of special events for the city of Chicago. She did all the big ones, at venues including Navy Pier, the Cultural Center, and Millennium Park. The latter's opening night had to be Donna's most impressive one. She pulled out all the stops for this lavish black tie event that brought together the crème de la crème of the city and state.

I was honored to even be at the event at a main table with lots of huge names. I ordered a drink from a passing waiter at the start. After he arrived with it, I was giving him a tip when he said, "No need for that, it's my pleasure. By the way, I'm the president of Target." He and I spent most of the evening talking to each other after that. I might have even brought him a drink.

Donna was really inventive in her work on so many of our film festival galas, including having actual taxicab meters on the gala dinner tables at the event we did for Jodie Foster. Remember her first big film was *Taxi Driver*. She's married to the famous TV anchor-journalist and narrator Bill Kurtis, whose rich voice is a major part of so many films and TV documentaries. They were mainstays with me at the film festival for so many years. I love them.

Candace Jordan was another memorable character during my festival years. She followed in the footsteps of other media personalities throughout the Sixties like Peg Zwecker, Maggie Daly, Virginia Kay, Pat Shelton, Gene Buck, Mary Frey, and Ann Gerber. Those ladies did social media before there was even a social media, which was really the newspapers of the day in the pre-internet era.

Candace was a Playmate of the Year over at Hugh Hefner's *Playboy* magazine as well as a bunny at the Playboy Clubs. I'm sure you can check her out for yourself at Playboy.com. She's proud of her *Playboy* past and has used it to great advantage. She is fearless with or without clothing. She and her retired ad executive husband are the toast of Chicago's social scene, a power couple and nice people, too.

Many of Candace's predecessors played the game with vicious nonstop gossip, some true, some not so true. Candace never did that. She always stayed positive, like the town crier without the crying.

Lynda O'Connor is a multiple award-winning book and author publicist who donated countless hours planning galas and PR events at the festival. She was the first on our team to get involved with foreign consulates working closely with the festival. She's also the one who convinced me to write this book and has been very important in my life.

Leslie Hindman is the star auction house maven. You know Sotheby's and Christie's? Well, she had Leslie Hindman Auctioneers in Chicago and across the country. She has been a close pal to me through thick and thin. In fact, she once decided we should get married. She pushed that idea for a year to everyone in Chicago: "Michael and I are getting married."

I mean, I know I'm a great catch, but she has two cats and I have saltwater fish, and never the two shall meet. We would have to live separately because of our pets. She knew I was pansexual (whatever that means). I figured she just needed a terrific "walker," but I took the whole thing seriously and on New Year's Eve one recent year, with my parents' engagement rings in my pocket, I went to her house and cooked my mother's famous recipe lasagna for 12 of her best friends. What I discovered was that these friends, some powerful ladies in the business, would be too difficult for me. They would never share her with me. That would have been too tough.

So serious or not, that effectively ended that.

The fact I wouldn't fit in with her social life became too clear to me, so I left that dinner at 10:30 p.m. Never mentioned the subject again. Put the rings away for another time and maybe another person. Yes, we are still friends.

I mean, I know I'm definitely a chick magnet, just not in, you know, the traditional sense. I loved it when a dear board member who later became the chairman of the festival, Jeanne Malkin, said to me one night while we were having dinner, "Michael, we should get married."

This was starting to become a regular thing.

"Jeanne," I said to my slightly tipsy dinner mate, "you're just lonely. You've had two terrific husbands and have three devoted children. I don't see it working. Second of all, I'm not Jewish."

"Well, true, there is *that*," she allowed, "and there is the other thing, too."

"Let's not go there," I replied. "If you need a walker, I'm good at that."

The subject never came up again. Jeanne was a wonderful, devoted friend of the festival and me. Evidently, I was seen as a prime marital asset for lonely older women. The truth is I was flattered, whether it was the alcohol thinking for them or not.

--

Linda Leemaster was the head of a marketing firm in Chicago. She specialized in focus groups and also was a Chicago Festival board member. In 1987, Linda unwittingly took on a new role. It was our opening night at the legendary Chicago Theatre, the U.S. premiere of Federico Fellini's latest film, *Intervista* (or *The Interview*) starring Anita Ekberg and Marcello Mastroianni.

We invited Ekberg. Alitalia Airlines gave us two first class tickets from Rome to Chicago. All was in place until Miss Ekberg arrived at Rome's airport and no tickets were there for her. It was 6 a.m. in Italy. She went ballistic. There was no Alitalia office to call to correct the situation and appease a very angry actress.

Back in Chicago, the show went on as scheduled. The film's producer was in attendance to speak to the press about it. The audience didn't ask where Anita was, as they saw her in the lobby and at the afterparty. They assumed she was too tired from the trip to go on stage. It didn't dawn on any of us that Linda Leemaster was a dead-ringer for Anita. Linda just went about her business helping guests at the theatre or at the party. She was unaware of the whole situation, as was I. We never heard from the real Anita Ekberg again.

--

When you viewed the festival from the inside, it was an incredible, creative, inspiring undertaking with young people from all sides of the city working together, being introduced to different cultures that they never would have been exposed to.

But you had to be a glutton for punishment to work for me. It was intense and chaotic, with long hours and never enough money to get things done the way they needed to get done. Some of our best things happened simply by hard work and luck.

Yes, I was the heartbeat of it all. You had to buy into my curiosity, enthusiasm, and optimism. You had to do it my way as it was my vision. It became yours as well. If it didn't, you moved on. This was not, after all, a democracy.

17 – ALL THE KING'S MEN

When I look back and think about those first few years of the Chicago Festival in the mid-Sixties, it's rather astonishing. It really took a great deal of vision and determination mixed in with my young naivete to convince people that my magnificent obsession could really play out successfully.

All I ever wanted to do was bring the world of international cinema to my hometown. Who knew you had to spend half your time proving to people that there was value to them in doing the festival? (Actually, maybe everybody.)

Once during a screening at Cannes, I remember speaking to an old friend named Richard Pena, the programmer of the New York Film Festival.

"Richard," I said, "I wish I could have your job of only going to film screenings day and night around the world, but I have to find the money too."

I had to prove to businesses like airlines, hotels and restaurants as well as city and state officials that the film festival had their market – the youth market, the older market, the gay market, the Chicago market, the International market. All of the markets. Was it worth their sponsorship, their money, their donation? They didn't care about the films. They wanted visibility and some return on their involvement/investment. It wasn't about the art for them but the commerce.

Movie stars tended to excite them more than my latest Italian or French films. A few names come to mind that changed the game.

Randy Crumpton, an African American lawyer and friend who showed me there was also a black audience, was a big key for us. He became a governing board member and knew how to navigate contracts, statutes, regulations, rulings, mandates, and decrees. Randy linked people in law, hip-hop, house music, film, and politics.

Through introductions and encouragement, Randy found us BIG money. We even developed an entire section of the film festival to showcase new, young black film directors and honor Hollywood's black luminaries like Pam Grier, Spike Lee, Morgan Freeman, Halle Berry, Sidney Poitier, Lawrence

Fishburne, and James Earl Jones. His favorite line was, "Everything is going to be fine, just take it easy."

I kept bugging Randy about his weight. I would say, "Randy, I could never be your pallbearer at this rate." I did end up being an honorary pallbearer as no way could I handle 400 pounds. Randy died much too young, at 53 in 2019.

A real surprise came from a film lover who attended our festival. This guy would see 20 or 25 films each year and loved them. He came to us one day and said, "I think I should be on your governing board." His name was Byron Pollock. He had retired from his manufacturing company in Chicago and wanted to be part of all the action. He was a big mover over at the Hubbard Street Dance Company but needed something new.

Byron got so involved with us that he was basically the behind-the-scenes chairman and financial advisor. He was an outspoken advocate for how he believed things should be done and was usually right, or to his mind he was ALWAYS right. But that was OK, we needed his smarts.

It's so rare to find a board member who actually *likes* films and can help run the show as well. Byron became hooked on Italian director Paolo Sorrentino's films, including *The Great Beauty, Il Divo,* and *The Consequences of Love*, and especially his actor Toni Servillo, who starred in all of them. I think it stemmed from the fact Byron resembled Toni and acted like him as well. He always wanted to meet Paolo and Toni on his many trips to Europe. I don't know if that ever happened before Byron passed away in 2019.

Who would think that a restaurant and its owner could have such a remarkable effect of the success of the Chicago International Film Festival? Joey Mondelli, a twelve-year-old who came to the United States from Bari, Italy, with his sister and older brother, many years later made it big in Chicago. Joey started the restaurant Kelly Mondelli. It had the politicians, the sports figures, the movie stars, you name it, including some mob bosses and me (a nobody using seven credit cards trying to pay for dinner). That's how I first met Joey.

Yes, after six cards didn't work, the seventh did, and we became lifelong friends. He knew I didn't have a pot to piss in but still introduced me to the politicians in Springfield, Illinois who could help me fund the festival. I sold them on tourism and how the festival helped the image of the state and filmmaking coming to shoot there.

Years have gone by, and Joey's many other restaurants now include La Scarola, a very popular hangout for all the cool people.

Back in 2017, one of those cool people was the acting legend Vanessa Redgrave, whom we were honoring at the festival and was married (and still is) to the Italian actor, producer, and director Franco Nero. We went to the restaurant. As I recall, it was a cold night, and Vanessa was in great form, very feisty and fun. I grabbed a corner table because I didn't want people to bother her.

Anyway, Vanessa ordered veal scallopini. We were relaxing with some wine and so forth, and the veal arrived. She took one look at it and declared, "This is not the way you make veal scallopini. Get somebody over here who I can talk to."

So, I screamed for the co-owner of the restaurant, Armando Vasquez, who was also the chef. Armando had no idea who Vanessa was. He just knew that he was getting an earful from this ballsy older woman.

"This is totally wrong," she informed him, and proceeded to add, "Here is how you make it." Vanessa then gave him a verbal lesson in preparing veal scallopini, which, to Armando's credit, he took in calmly even though he had to be hugely offended.

He came back with the dish, but Vanessa insisted, "This is still not right." She continued to teach him the proper way to prepare what is clearly a signature dish of this highly respected Italian establishment. Finally, Vanessa was satisfied that it was perfect, and she was happy with it. But oh my God, I thought we were going to go to war right there at the table.

It was so funny. I remember Armando after that taking me aside when Vanessa was out of earshot and telling me, "Don't you ever bring that lady back in here again." And once more, he had not the foggiest clue he had just been taken to school by the great Oscar® winner Vanessa Redgrave – making friends with her helpful food prep suggestions.

--

I also reflect back on our gala tribute to Clint Eastwood in 2002 out at Navy Pier with around 800 guests when something rather amusing happened. It was my job, as always, to keep the honoree free from the guests pestering him for photos and autographs and make the show run smoothly. Clint spotted a face in the crowd, one that I had kept away, as his wife wanted a photo with Clint.

"Who is that guy with the great face?" Clint asked. "I want to meet him."

"Oh, he has this hole-in-the-wall restaurant," I replied, to which Clint said, "Ya never know when you'll need a good hole-in-the-wall restaurant, bring him over!"

The wife got the photo, and Joey Mondelli met the man.

Speaking of Clint Eastwood, he and his entire family so enjoyed the new hotel where we put them up that they actually came back and paid for it the next time. It was the new Park Hyatt Hotel, and the suite was amazing, occupying what seemed like half a floor overlooking the Chicago Water Tower.

It was made possible by a really smart guy, Ed Rabin, who was on our governing board for five years and the president of Hyatt Hotel Corporation for many, many years. He was our treasurer, and he understood how sponsorship worked more than most. For years, we kept our top celebrities at the Park Hyatt. Ed and his wife Jackie remain close friends. He always believed in me and also forced me to write this book.

--

What do you do if you're a successful lawyer and real estate bigshot in Chicago and you get restless? You get involved with a buddy who has a film company and make it a big success story.

Tom Rosenberg is that guy. In 1988, we honored British director Alan Parker, who did some revolutionary films back then, including *Midnight Express, Mississippi Burning,* and *Fame.* He also made some fabulous television commercials early in his career. Years later, I would be asking Tom if we could premiere Alan's latest film that he'd produced, 1991's *The Commitments*. It turned out to be that remarkable film's success that got Tom hooked in the business. Money will do that

Rosenberg formed Lakeshore Entertainment in 1994 with Ted Tannebaum and other Chicago buddies, but Tom was the brains. He won an Oscar® in 2005 with *Million Dollar Baby,* the Eastwood flick, and his dark 2003 film

Underworld became a major franchise for Lakeshore. Tom gave us a few films to open the festival over the years, including *The Human Stain* in 2003.

But it was *The Stand-Up Guys* that we world premiered in 2012 that should have been his next Oscar®. It didn't work, I felt, because Tom allowed a young director (Fisher Stevens) who had never done a feature before take on the project. With three heavyweight actors in Al Pacino, Alan Arkin, and Christopher Walken, and no seasoned leader, the film failed.

Tom is a good guy. He's produced over 70 features. Lakeshore Entertainment is gone now, but their film library sold for lots of money. I wonder what he'll do next.

--

It's funny how it all came together as sponsors and political support changed with the times. Various Chicago mayors became enthused about the international flavor the festival brought to the city.

Mayor Jane Byrne was the best. She felt the festival fit into the idea of Chicago as "a world class city." We even created a Movies in the Parks initiative for her back then. When city administrations began to ignore the festival later on, Governor Jim Thompson saw the importance, understood The Big Picture, and saw the international image of the city and the state raised. It served to fund the festival for many years. My magnificent obsession was able to continue well past the new millennium.

Actor Willem Dafoe enjoys a light moment at a Chicago
International Film Festival tribute. (Robert Dowey)

Oscar®-winning actress Viola Davis (right) and husband Julius
Tennon being photographed at our festival. (Robert Dowey)

The luminous Oscar®-nominated actress Uma Thurman. (Robert Dowey)

The Oscar®-winning producer-director Taylor Hackford, who also happens to be Helen Mirren's husband. (Timothy M. Schmidt)

Oscar®-winning actress Rachel Weisz. (Robert Dowey)

Chicago-based TV entertainment reporter and film critic Jake Hamilton (right) interviews Patrick Stewart at our festival. (Timothy M. Schmidt)

Actor Oscar Isaac ponders a question at
the Chicago Festival. (Timothy M. Schmidt)

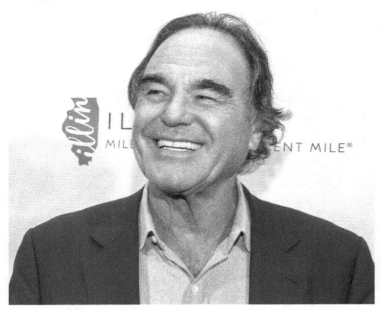

Writer-director Oliver Stone, who has won three Academy Awards®,
is all smiles at the festival. (Timothy M. Schmidt)

At the premiere of the movie *Marshall* (the story of Supreme Court Justice Thurgood Marshall) in 2017. (front, left to right) Co-star Marina Squerciati, me, co-star Jussie Smollett; (middle, left to right) producer Paula Wagner, star Chadwick Boseman, Thurgood Marshall's son John Marshall; (back row, left to right) co-star Sterling K. Brown, producer-director Reginald Hudlin. (Timothy M. Schmidt)

Oscar® nominee and Emmy winner Laura Linney (who has the greatest smile, at left) and me (who doesn't). (Timothy M. Schmidt)

Oscar® winner Kevin Kline, who I found to be an incredibly
nice and funny guy. (Timothy M. Schmidt)

The Academy Award®-nominated character actor John C. Reilly, who showed at our
Chicago Festival that he knows how to wear a hat. (Timothy M. Schmidt)

The young, up-and-coming indie film director
Joe Swanberg. (Timothy M. Schmidt)

The legendary cinematographer Haskell Wexler, a two-time Oscar® winner, graced
our festival with his presence. (Timothy M. Schmidt)

The actor, producer, and director Fisher Stevens. (Timothy M. Schmidt)

Talk about a film festival with an international flavor. (left to right)
The Turkish-Italian director Ferzan Ozpetek, the American actress (and friend)
Kathleen Turner, Iranian director Parviz Shahbazi, the German actor-director
Margarethe von Trotta, the Israeli-born cinematographer Giora Bejach –
and me in front in the middle. (Timothy M. Schmidt)

The Chicago International Film Festival audience is always
up for a good time. (Timothy M. Schmidt)

The Broadway legend Elaine Stritch (in hat) addresses the festival crowd in 2014.
She died just a few months later. (Timothy M. Schmidt)

Me and the two-time Oscar®-winning Mexican director Guillermo
del Toro (left), posing for the ultimate selfie. (Michael Kutza Archives)

Irish actor Colin Farrell (left) and Norwegian acting and
directing legend Liv Ullmann were a big presence at our
50th Chicago Festival. (Timothy M. Schmidt)

Walkens are always welcome at our festival, particularly when they're Oscar® winners named Christopher. (Robert Dowey)

Director Robert Zemeckis, born in Chicago and an Academy Award® winner for *Forrest Gump*. (Timothy M. Schmidt)

The Oscar®-winning brothers Ben (left) and Casey Affleck. (Robert Dowey)

Al Pacino, winner of an Academy Award® for *Scent of a Woman* in 1992, grins his way around our festival. (Robert Dowey)

Me at left with La Scarola chef and co-owner Joey Mondelli in his legendary Italian eatery. (La Scarola Staff)

(left to right) Liv Ullmann and Colin Farrell cut the cake on our Chicago International Film Festival 50th Anniversary Celebration in 2014 along with me and the amazing cake's creator, Maureen Schulman. (Timothy M. Schmidt)

The phenomenal German-American architect
Helmut Jahn. (Timothy M. Schmidt)

Managing director Vivian Teng (left) and artistic director
Mimi Plauchè, co-heads of the organization who have brought
great vitality to the Chicago Festival. (Dan Hannula)

The Oscar®-winning documentarian and all-around
rabble-rouser Michael Moore. (Timothy M. Schmidt)

Actor Robert Downey Jr., looking stylish and having a
great time in Chicago. (Timothy M. Schmidt)

Film producer Gigi Pritzker (in the middle) in 2010. (Timothy M. Schmidt)

Oscar®-nominated director Peter Bogdanovich, who presided over many memorable films including 1971's *The Last Picture Show.* (Timothy M. Schmidt)

Kicking back with a lively group of guest directors at the Chicago Festival. (Timothy M. Schmidt)

Standing between the French director Claude Lelouch and actress Geraldine Chaplin, daughter of Charlie Chaplin and a dear friend. (Dan Hannula)

Lelouch and Chaplin having an emotional exchange at the festival. I wish I knew what they were discussing. (Dan Hannula)

A beaming Jane Fonda (a two-time Oscar® winner) at her
Chicago Festival Tribute in 2017. (Timothy M. Schmidt)

Oscar®-winning singer and actress Jennifer Hudson cuts
an elegant figure at our festival. (Timothy M. Schmidt)

Mexican actor and filmmaker Alfonso Arau. (Timothy M. Schmidt)

The now-retired Chinese/Hong Kong filmmaker known as Scud (center) has an obvious propensity for the provocative. (Artwalker Ltd)

Filmmaker Andrew Davis clearly never heard that it's not polite to point in public, but he's a great director and friend. (Timothy M. Schmidt)

With my dear friend, the brilliant producer
Paula Wagner (left). (Timothy M. Schmidt)

Honoring the Academy Award®-winning couple Taylor Hackford (left) and his luminous actress wife Helen Mirren at our 2017 Chicago Festival. (Timothy M. Schmidt)

18 – THE FILM CRITIC

From the very start, I knew I could never pull off doing stage interviews with movie stars or film directors myself. I'm just too shy, even though that might surprise people who meet me and see how naturally gregarious I am.

Nonetheless, I enlisted film critics and various film authorities to do the job at the Chicago Festival. I knew I could discover the films for the festival. That was really my primary job. How? I had created contacts at film institutes, embassy cultural departments, and with a string of film authorities around the world. I think I was initially welcomed by these folks because I was young and from Chicago, the home of Al Capone and Mayor Daley – and the world's greatest pizza.

As alluded to earlier, I asked various film people, including John Russell Taylor, the film critic from *The Times of London*, David Robinson of *The Financial Times*, and John Kobal, a delightful, slightly crazy, London-based film historian and scholar who knew everybody. I put them in the hot seat in Chicago to cover for me and my bashful self. The audiences loved them.

Looking back, it's clear that I had a tendency to irritate the local press just because of the way I came across. I didn't really care. I guess they all resented me from year one of the festival. How dare I even attempt this event and choose the films myself? Who the hell did I think I was, anyway?

Yes, I was the young prodigy back then. I was hated by the press but adored by the public, or at least the film festivalgoers. You can't satisfy all of the people all of the time, so I stopped trying and just had a good time, focusing on my strengths and feeling superior to those trying to discredit me and my creation.

It took years, but I learned that I should also start using some local film people who I felt I could sort of trust. Never trust a journalist, Colleen Moore always said. I learned that particular lesson over and over again.

In the early years, I utilized brilliant film students from the University of Chicago film organization Doc Film. I invited Charles Flynn to introduce George Cukor. He also pulled the film clips, including mistakenly pulling a clip from the wrong version of *Little Women* that George didn't direct. It was embarrassing, but what are you gonna do? Shit happens.

I didn't use Charles again, however.

The great Roger Ebert of the *Chicago Sun-Times* was the new kid in town early in the festival's life. His first job out of college was to cover the Chicago Festival in 1967. I liked Roger and invited him to his first two international festivals: Tehran Film Festival in Iran, as I was working on that festival for the Shah of Iran, and then I invited him to the Venice Festival. He was becoming known on the international circuit.

After that, I started inviting Roger on stage to do the intros, which he performed brilliantly. The man did his homework and loved the stars like Kim Novak and Morgan Freeman, and directors like Sydney Pollack and Russ Meyer. He was smart enough to continue a relationship with most of them. He even scripted the really terrible, campy *Beyond the Valley of the Dolls* and the equally campy *Beneath the Valley of the Ultra-Vixens* for Meyer. I never stayed chummy like that with the honorees, as I just wanted to move on to the next big discovery or tribute. It worked out better for me that way.

There was a drawback with Roger in the early years. He was a drinker and required us to take the visiting directors to meet him at his favorite bar late at night. That bar was O'Rourke's Irish Pub. As many of our guests pointed out over the years, Roger was always at the bar, drunk with his cronies, and we felt we were taken there essentially just to kiss his ring. Many years later, Roger did go to Alcoholics Anonymous to resolve his drinking problem.

However, Roger was a fan of the festival and went out of his way to help put us on the map and give us legitimacy in a way no other critic could — that is, until he fell in love with the Toronto Film Festival and all of its gold-plated perks. But it was AA that straightened him out along with a young lady named Chaz (whom he met in AA) with a ready-made family that totally changed Roger. He ultimately became the first film critic to win the Pulitzer Prize for Criticism. And he deserved it.

Chaz basically saved Roger's life. We'll always be grateful to her and love her for that.

Roger was often quoted as saying, "It was at the Chicago Festival that I met all the future giants of cinema like Scorsese, Fassbinder, Tavernier,

Varda and Greg Nava. The festival is an adornment to our city." Thank you, Roger.

Gene Siskel from the *Chicago Tribune*, with whom Roger was paired on television for many years, was a sportswriter who later switched to film reviewing. But he could be snarky and too often unprepared. It took him many years to warm to our festival. I had him introduce an evening with director Bill Friedkin (*The French Connection, The Exorcist*), and Gene became a fan after that.

Gene had been a sportswriter before becoming a movie critic. I always thought he came at film criticism from a truck driver's/blue collar approach, whereas Roger was in every way artistic and very much understood the film world and had great respect for the filmmakers.

Before Siskel and Ebert became household names as *Siskel & Ebert at the Movies*, they both would of course cover our Chicago Festival film screenings as part of their jobs at their respective newspapers. I remember a few funny moments in the early years. Gene tended to fall asleep at screenings, while Roger was always eating something like giant sandwiches.

We had the world premiere in 1978 of John Carpenter's *Halloween* at midnight. Gene slept through the whole film and yet the next day gave it four stars. But my greatest memory was when Gene woke up midway through a South Korean film we had entitled *Castle of the Rose* and asked Roger, who was seated next to him, "Did I miss anything?"

Roger always loved telling this story. "Well, Gene," he told his newspaper competitor, "the star of the film had sex with her French Poodle." Gene didn't sleep through that many films after that at the festival.

I really put Russell and Kobal through their paces. I remember when John was interviewing director Vincente Minnelli, the man who directed more than 30 musicals for MGM (including *An American in Paris, The Band Wagon,* and *Gigi*) when Minnelli suddenly went blank on stage at the Granada Theatre. He had no idea where he was or even why he was there. John covered beautifully, since he knew Vicente so well, and he did the entire interview by himself. He asked the questions and then answered them. He was brilliant.

Kobal found himself and actress-dancer Ann Miller (*On the Town, Kiss Me Kate, Easter Parade*) on stage with a very small audience, maybe 400 in a theatre of 3,000 seats. He said to the festival staff, "Just keep shining a spotlight at Ann's face, so she can't see the nearly empty house." It worked beautifully.

When I'd first met Ann at her Beverly Hills home years earlier, I recall her having said, "Oh, I have to go now and become Ann Miller." I had only heard anything like that one time before, when we did the tribute to Ann-Margret. We had finished lunch at her hotel and her husband Roger Smith said, "Ann, it's time for you to become Ann-Margret."

I guess that's showbiz.

Kobal was also faced once with a full house and Gloria Swanson on stage, and an audience that only wanted him to grill her about her affair with Joseph P. Kennedy and then her diet but nothing about her films. As Kobal said, "God, there seem to be more food freaks than movie freaks here tonight." It was incredibly strange. You have *Sunset Boulevard* right in front of you, and they want to know about tofu.

Robinson came in to do an evening with animator Richard Williams, famed for *Who Framed Roger Rabbit?*. It was hysterical when Richard described his endless conversations about the bust size of Jessica Rabbit, voiced by Kathleen Turner. The director, Robert Zemeckis, wanted the breasts larger, while Steven Spielberg, who produced the film, wanted them still larger, and the waist thinner. For an animated rabbit!

Russell's unforgettable evening with MGM's noted costume designer Helen Rose (*Butterfield 8, Cat on a Hot Tin Roof, Silk Stockings, High Society, Ziegfeld Follies*) happened at the Biograph Theater with a full house. They both went up to the raised stage platform in front of the movie screen and sat in their director's chairs, John in his and Helen in hers, except that her chair was a bit further back on the platform. Consequently, when she sat down, it seemed the back legs of the chair slid too far back, right off the edge of the stage.

Helen flew backwards and completely disappeared. All you heard was a loud gasp from the 750 people in the audience. She then got up and John helped her climb back on stage as if it had never happened. I can still see it and feel it and hear that gasp, even today.

When it was over, I said to John, "You were nervous. I've never seen you nervous before." He replied rather irritably, "You'd be nervous too if your *seventysomething* star had just crashed *arse* over tip into the outer darkness." I guess it was a tribute to the consummate professionalism of them both that for the moment, I had completely forgotten.

There was another potential drama that night. An audience member turned up who was such a fat, unkempt, poorly dressed, probably drunken slob that the ushers wanted to eject him, lest he prove a troublemaker. But he stayed quiet until the Q&A, when he got up and asked, in the daintiest possible tones, "Mith Rothe, that *fabulous* dress Cyd Charisse wears in *Deep in My Heart*, is it spun silk or tulle?"

It was years later that I received a call from Helen's daughter asking if I would do a tribute to her mother and raise money so she (the daughter) could have trans surgery. I said we had already done a tribute to Helen. "My name is Joe and I need the money," she declared. I remembered once meeting the daughter back in the day and had thought nothing of her having a corncob in her blue jeans. To me, that was just a tomboy, as we would have called her back then.

I realized a little later that things were a bit more complicated than that. Apparently, the daughter had resented being a girl from an early age, which was especially unfortunate considering her mother was famed as the designer of all the most gorgeous dresses that made Elizabeth Taylor a top Hollywood glamour star. Her annoyance at her lot extended to the point that, when Helen designed something drop-dead glamorous for her graduation prom, in a fury the girl drove out to the Valley and dropped it unceremoniously in a trash bin in some anonymous shopping mall car park. I've always wondered what lucky young woman found and wore it.

Anyway, I wished her luck in her quest for sex change money, saying it was just not possible to do another tribute to her mom while pondering life's little ironies. I never heard from her again.

--

Rex Reed, the New York actor, film critic, and TV host, was brought in for a few interviews, his best being with Ann-Margret. Rex knew Ann well and could get anything out of her, including asking about her love affair with Elvis Presley, which destroyed Ann on stage into a crying jag. Rex also questioned Angela Lansbury and Neil Simon brilliantly. But the giant love-in between Rex and the audience was with Angela.

I invited a local TV entertainment reporter, Dean Richards of WGN-TV, to do intros a few times. He could be great, or incredibly preoccupied with his next assignment that day. He's a perfect example of an overextended reporter. He delivered with an edge that made audiences uncomfortable and always seemed bitter about something. But with age, he seems to have mellowed.

Then there is a dear, sometimes close friend, Bill Zwecker of *Fox News* and *Chicago Sun-Times* journalistic fame, who did many intros for us. He was always prepared, but never comfortable or believable in his delivery. I recall Bill being really excited when he scooped the other papers in town about my possibly being fired from the festival. He said, "This is bigger than when I scooped everyone telling the world that gold medal Olympic diver Greg Louganis had AIDS."

Bill made the front page of the newspaper, as did I. But Bill, like Dean when he is on camera, sometimes seemed passionless, as if he were in a hurry to go to another engagement, another fabulous film junket in L.A. or Paris. Which was probably the case. The lesson here is to never get ahead of yourself and always stay in the moment. Otherwise, too many can tell you're multitasking.

I never knew if Bill got tired of covering the entertainment beat or if Fox32 got tired of Bill. Like everything else in the media business, there is always someone waiting in the wings. In this case, it was a very clever young man who was a film critic in Texas at the tender age of 14 and later moved to Chicago. Jake Hamilton worked alongside Zwecker for a while and then, in an *All About Eve* moment, took on the key role. At 30, he already has an Emmy.

--

Richard Roeper, a gifted young writer at the *Chicago Sun-Times* and Gene Siskel's successor on *At the Movies* alongside Roger, did an intro at our Robin Williams tribute but made the grievous error of trying to be funnier that Robin. It did not go over well. Now don't get me wrong, Richard is a brilliant guy and very funny on the page. But as a stand-up comic, not so great. It's all got to be in the delivery.

It was years later that Roeper did our tribute to Jane Fonda and was brilliant. He is in many ways our new Mike Royko, the famed newspaper columnist legend in Chicago. Richard is on top of his game as I write these words.

Then there were the really negative film personalities whom I just couldn't abide at the Chicago Film Festival as long as I was running the show. The ever-pompous Jonathan Rosenbaum writes for a little local newspaper called *The Chicago Reader*, our version of *The Village Voice* in NYC, an arts/cultural neighborhood sheet.

Jonathan was a world authority it seemed on Orson Welles and Iranian cinema. His yearly reporting on our festival consisted of, "Why isn't it the Toronto Film Festival?" He would write the same story year after year: "Why does it not have the same 200 films that I liked in Toronto?" What can I say? The guy is respected by his followers, but certainly not my cup of tea.

He reminded me of onetime *Chicago Tribune* film critic Michael Wilmington. Brilliant guy, very laid back, studious yet loveable. He was devoted to his 92-year-old mother, but still found time for reviews and doing intros at the festival. I noticed that as he became more recognized, he would voice a rather strong opinion about our film selection: "There are 55 films in this festival that I have never heard of. They can't be any good."

I liked Michael. I only wished he'd liked himself a bit more. He could have gone so much further.

--

The majority of film critics in the world are, well, slobs. They live in the dark and really don't come out into the light very much, so the need to dress or shower doesn't seem important to them. Seriously, as a group they're anti-social and anti-shower.

However, I can think of two in the world for whom this wasn't the case. Alexander Walker was with *The Evening Standard* in London. He was a suit and tie guy, elegantly put together at all times even through multiple hours and films. A class act all the way. I recall one year that I had invited him to be on our prestigious feature film jury. British Airways canceled his London to

Chicago flight for some reason, and he went ballistic at London's Heathrow Airport, explaining to the British Airlines staff how important he was.

"This Chicago event opens tonight, and I am the president of the event!" he fumed, demanding that he be put on the Concorde. They did, and he was on time in Chicago. Class and style and a loud voice when he needed it. That was Alex.

The other best-dressed film critic lives in Chicago: Michael Phillips of the *Tribune*. He took on the job after Siskel died. Phillips was the drama critic first at the *Tribune, Los Angeles Times,* the *St. Paul Pioneer Press,* and the *Dallas Times Herald.* I am assuming he had to dress appropriately to attend live theatre, and it just carried over. He has a cool intellectual style about him and presents well. He's now quite definitely the best in town and well respected on the festival circuit. I also like him as a person. Consider me a fan.

All of these film professionals came forward time and time again to help me and the festival. They gave the organization the great credibility we needed. So, why didn't I just do the intros myself? Today I'll do some, with lots of preparation, so I look unrehearsed and casual. But I admit to still being rather insecure on stage. I have never been satisfied with festival staff members doing these things. No matter how well versed they are, they just are never comfortable or charismatic.

I believe that the Chicago Festival audience should have the best qualified person there to do the interviews of stars and directors at all events. It must be a special event for them or there's just no point.

19 – CHICAGO

At the turn of the 20th century, Chicago was a pretty big deal in the moviemaking business, much as was New York and (later) Hollywood. It was founded as the Essanay Film Manufacturing Company and later Essanay Studios.

It's not an exciting name, but it had some exciting names on the lot. They included Gloria Swanson, Wallace Beery, Ben Turpin, Francis X. Bushman and even Charlie Chaplin. Broncho Billy westerns were developed there but shot in California. Louella Parsons, later the famous gossip columnist in Hollywood, was a screenwriter there. Colleen Moore did her first screen test there, and it was then sent to Hollywood for D.W. Griffith to see and later give her a contact as a favor to her uncle.

Outrageous as it may seem, Swanson married Beery on the backlot of Essanay in 1917. She was only 16 ½ years old. The place cranked out two or three films a week to meet the demands of nickelodeons, but eventually most productions moved to better weather and studios in California.

Chicago was continually seen in the films in the Forties. I wonder how many times our Union Station was used in a film. We all lived through the Mayor Daley years, when he stopped all production because he was so pissed at Haskell Wexler's *Medium Cool* in 1969. The good mayor evidently didn't like seeing himself on the world stage, over and over again, saying, "Shoot to kill!" during the notorious 1968 Democratic National Convention. He got over it after a few years while insisting on having script approval over anything shot in town for a while.

With the Chicago International Film Festival, we continually promoted coming to Chicago to make your next film, working closely with the various film offices and at the Cannes or Berlin festivals, hosting receptions heralding the city and the state. My purpose in starting the festival was, through film, to show my hometown and what they were missing to the world. It certainly succeeded in doing that.

It was always important to me to present the newest and best local productions and feature them at the festival. It's the perfect showcase to roll out their new work. I think of director Joe Swanberg's first feature and future films all being presented. Joe was just a young man working in our office handling travel and guests. Today, he's a terrific director. I'm not saying that some of the magic must have rubbed off just being near all these famous folks, but it can't help but encourage a young filmmaker.

In the Eighties, it became a big deal to make your movie in Chicago. Some of my favorites that really show off the city are *The Untouchables, The Blues Brothers, Ordinary People, Risky Business, The Fugitive,* and all of the John Hughes films, especially *Ferris Bueller's Day Off.* Chicago is and will forever be Gotham City, the home of Batman. *The Dark Knight* director Christopher Nolan really understood the city and its landscapes, even shooting in the IMAX system.

They always say you can't become a star in your own hometown. You have to get out, like Michael Mann, Andrew Davis, Robert Zemeckis, Haskell Wexler, Marlon Brando, Bob Fosse, Bill Friedkin, Charlton Heston, Vincente Minnelli, Christopher Nolan, Harold Ramis, Mark Romanek, Orson Welles . . .the list goes on and on.

It seems that TV series today are pretty much singlehandedly keeping the year-round filmmaking business alive. We finally have a giant film studio, Cinespace, right here in Chicago. Hollywood will always come and shoot exteriors here, while interiors will be captured back in New York or Los Angeles.

I think the saddest thing was to have the 2002 Oscar®-winning film version of *Chicago* shot in Toronto. That really sucked.

20 – A HIGHLIGHT CHRONOLOGY FROM THE FIRST HALF CENTURY-PLUS YEARS OF THE CHICAGO INTERNATIONAL FILM FESTIVAL

1964 Michael Kutza, age 22 (but passed for 27), is determined that Chicago should be a home for the appreciation of international film. Legendary *Chicago Sun-Times* columnist **Irv Kupcinet** is an early supporter and introduces him to the recently widowed silent screen comedienne **Colleen Moore Hargrave**, then living in Chicago, who opens many doors for Kutza among Hollywood elite and Chicago society. Kutza founds Cinema/Chicago, the organization that presents the Chicago International Film Festival.

1965 The Chicago International Film Festival debuts at the original Carnegie Theatre at Rush and Oak streets on November 9. Eight films are shown in the Feature category, chosen from more than 300 entries from fifteen nations. Other categories include Religious Films, Experimental Films, Industrial Films, Short Subjects, Educational Films, Television Commercials, Documentaries, and Cartoons. In its inaugural year, the festival honors **King Vidor** in a ceremony attended by **Bette Davis**, and **Stanley Kramer** gives a public talk, offering advice to student filmmakers. The first Gold Hugo for Best Feature goes to *The Lollipop Cover*. Short films by the soon-to-be-legendary director **William Friedkin** are presented.

1966 Kutza announces that the festival is open to "Adults Only" to avoid having to obtain approval for his programming from the Chicago Censor Board, a civic group made up of policemen's widows that rated films for local audiences. The board is ultimately dissolved by the U.S. Supreme Court two years later. **Harold Lloyd**, **Otto Preminger**, and **Colleen Moore Hargrave** are celebrated at the festival.

1967 The festival presents *I Call First* (later renamed *Who's That Knocking at My Door*), the first feature film by **Martin Scorsese**, starring **Harvey Keitel**. Scorsese's film is reviewed by **Roger Ebert** as one of his first assignments for the *Chicago Sun-Times*. He writes that the film "made a stunning impact in its world premiere Wednesday night at the Chicago International Film Festival." Honored are **Busby Berkeley**, **George Cukor**, **Ruby Keeler**, graphic designer and filmmaker **Saul Bass**, **Pablo Ferro**, and animator **Richard Williams**. The festival's iconic "eyes" logo, a composite of close ups of silent screen sirens **Theda Bara**, **Pola Negri**, and **Mae Murray** and designed by Kutza, debuts.

1968 Two short films by **John Lennon** and **Yoko Ono**, *Two Virgins* and *No. 5*, receive their world premieres at the festival. The festival pays tribute to filmmaker **Mervyn LeRoy** and famed Scottish-Canadian animator **Norman McLaren**. The winner of the Gold Hugo is *Innocence Unprotected* from Yugoslavian director **Dusan Makavejev**.

1969 The Festival debuts a New Directors series as well as a Films for Children competition judged by children. For the first time, the festival receives public acknowledgement from **Mayor Richard J. Daley**. The mayor had earlier told Kutza, "The films you show could lose me votes!" and declined public support of Kutza's audacious, often experimental, film programming. Mayor Daley feared it could prove a political liability, though the administration had quietly helped the festival to secure hotel partners and theaters. The career of Hungarian-American animator **George Pal** is celebrated. Belgian director **Harry Kümel** is awarded the Best Feature Gold Hugo for *Monsieur Hawarden*.

1970 Illinois Governor Richard B. Ogilvie publicly supports the festival, writing, "Too often, the avant-garde image of the filmmaker has been interpreted as antithetical to the mid-American ethic. But that is a view which disregards an essential element of that ethic: its firm foundation in the concept of individual freedom. Film is free, as America is free." Fifteen years of grants from the State of Illinois follow. The festival features its first all-critics jury, with banter and barbs exchanged by the *Chicago Sun-Times's* **Roger**

Ebert, the *Chicago Tribune's* **Gene Siskel**, *Chicago Today's* **Mary Knoblauch** and the *Chicago Daily News's* **Sam Lesner**. **Howard Hawks** and **George Stevens** visit the festival and discuss their work and influences. Celebrated documentarian **Les Blank's** short *The Blues Accordin' to Lightnin' Hopkins* takes home a Gold Hugo.

1971 Film Festival honorees include figure skating world champion and film star Sonja Henie, Hollywood director **Franklin Schaffner**, and film/television producer-director **Donald Siegel**. Polish director **Krzysztof Zanussi's** *Family Life* and **Claude Jutra's** *Mon Oncle Antoine* are among the films that debut at the festival, with Zanussi's film winning the Special Jury Prize and Jutra's taking home the Gold Hugo for Best Film. **José Luis López Vázquez** wins the Best Actor honor for *The Ancines Woods.*

1972 Grigory Kozintsev's *King Lear*, representing the USSR in the festival, is among the notable films screened. Film festival honorees include renowned documentary director **Frederick Wiseman**, featuring a retrospective of his work, as well as French filmmaker **Abel Gance**, **Paul Morrissey**, and **Linwood G. Dunn**. For the second consecutive year, **José Luis López Vázquez** is honored with a Best Actor Silver Hugo, this time for his performance in *Mi Querida Señorita*. British director **Mike Leigh's** directorial debut *Bleak Moments* wins the Gold Hugo for Best Film.

1973 Film festival highlights include a complete retrospective of the work of Indian auteur **Satyajit Ray** and tributes to **David L. Wolper** and classic Warner Bros. musicals. **Rainer Werner Fassbinder's** *The Bitter Tears of Petra Von Kant* wins the Special Jury Prize. Greek master **Theo Angelopolous's** first feature *Reconstruction* is shown along with his *Days of '36.*

1974 Festival honorees include **Robert Wise**, **Vincente Minnelli**, Slovak film writer and director **Ján Kadár**, and **Angela Lansbury**. The Gold Hugo for Best Feature goes to **Giorgi Shengelaia's** *Prosmani* from the USSR. *The Clockmaker of St. Paul*, the feature debut of French filmmaker **Bertrand Tavernier**, also has its premiere at the festival.

1975 Milos Forman's classic *One Flew Over the Cuckoo's Nest* has its world premiere during the opening night of the festival with **Jack Nicholson** and other cast members in attendance. A tribute to **Pier Paolo Pasolini** is planned, but the celebration becomes a memorial when the director is murdered the night before his departure for Chicago. Also honored is director **Stanley Donen.** Notable screenings include Polish filmmaker **Andrzej Wajda's** *Land of Promise*, which takes home the Gold Hugo.

1976 Notable debuts include the **Albert and David Maysles'** cult classic documentary *Grey Gardens* and **Wim Wenders'** *Kings of the Road*, which is awarded the Gold Hugo. Native Chicagoan **Charlton Heston** and Metro Goldwyn Mayer costume designer **Helen Rose** are honored.

1977 The U.S. premiere of **Peter Weir's** *Picnic at Hanging Rock* and **Sidney Lumet's** *Equus* debut at the festival. It also features a retrospective of **Mel Brooks'** films and tributes to **Ann Miller, Lindsay Anderson, Krzysztof Zanussi**, and costume designer **Edith Head.** Celebrated Cuban auteur **Tomas Gutierrez Alea's** *The Last Supper* wins the Special Jury Prize.

1978 President Jimmy Carter offers the festival his congratulations. The fest presents the world premiere of **Martin Rosen's** much anticipated animated *Watership Down*. Honorees include **Orson Welles** (who, unable to attend, creates a special short film for the occasion) and **Mickey Mouse**, who arrives from Hollywood via train and meets with **Chicago Mayor Michael Anthony Bilandic** before a screening of clips from classic Disney animation. The Hungarian film *A Quite Ordinary Life*, directed by **Imre Gyöngyössy** and **Barna Kabay**, and Spanish director **Jaime Chavarri's** *To an Unknown God* debut at the festival.

1979 The Chicago Theatre hosts the opening night of the festival for the first time, thanks to support from **Mayor Jane Byrne**. Special effects master **Peter Ellenshaw Sr., Karen Black**, and **Maximillian Schell** receive honors, and films from 24 nations are screened. A festival favorite is **Victor Nuñez's** *Gal Young 'Un*, and the Gold Hugo for Best Feature goes to *Angi Vera*, a Hungarian drama from director **Pál Gábor**.

1980 Taylor Hackford's first feature film, *The Idolmaker*, opens the festival. A special selection of new films from Scandinavia is screened, and **John Houseman**, **Gloria Swanson**, **Neil Simon**, **William Friedkin**, and film editor **Verna Fields** are honored. Program highlights include the Polish film *Camera Buff*, the first feature from director **Krzysztof Kieslowski**, and German filmmaker **Percy Adlon's** debut *Celeste*.

1981 Dedicated to and attended by French New Wave master **François Truffaut**, the festival features a 12-hour marathon of the filmmaker's work. Opening night features a presentation of a restored print of the 1924 film *Peter Pan* at the Granada Theatre, accompanied by a full orchestra from the Eastman House. Also honored are Argentine film director **Leopoldo Torre Nilsson** and *Looney Tunes* animator **Isadore "Friz" Freleng**. New German Cinema leader **Margarethe von Trotta's** *The German Sisters* (alternatively titled *Marianne & Juliane*) wins the Gold Hugo. The festival presents **Peter Greenaway's** directorial debut, *The Falls*.

1982 President Ronald Reagan praises the festival, writing, "The Chicago International Film Festival serves both artists and the public by providing a forum for discussion and sharing." The festival presents a tribute to the recently departed **Rainer Werner Fassbinder**, featuring five of his films. **Ann-Margret** is on hand to speak to festival guests and discuss her career. **Robert Altman's** *Come Back to the Five and Dime, Jimmy Dean, Jimmy Dean* takes home top honors, and Altman makes his first of many appearances at the festival.

1983 The festival showcases a retrospective of early British musicals, a series of both French and Spanish films, and a tribute to **Jane Russell** and also welcomes *Rocky & Bullwinkle* cartoonist **Jay Ward**. French New Wave auteur **Jacques Demy** participates on the international features jury. Australian filmmaker **Paul Cox's** *Man of Flowers* is screened, and Spanish director **Victor Erice's** *The South* takes home the Gold Hugo for Best Feature.

1984 The 20th Anniversary festival salutes Italian comedies and classic monster movies as well as special effects innovator **Douglas Trumbull** of

2001: A Space Odyssey and *Blade Runner* fame. The Gold Hugo for Best Film goes to *Khandhar*, directed by **Mrinal Sen** from India, while Danish director **Lars von Trier's** *The Element of Crime* wins the Silver Hugo. Program highlights include **Leos Carax's** debut film *Boy Meets Girl* and **Mike Leigh's** *Meantime*.

1985 Dancing and international intrigue open the festival with the premiere of Hackford's *White Nights*, starring **Mikhail Baryshnikov**, **Gregory Hines**, and **Isabella Rossellini**, who are all in attendance. *Bonnie and Clyde* director **Arthur Penn** and camp icon **Russ Meyer** are celebrated. Argentine director **Luis Puenzo's** *The Official Story* has its debut in Chicago before going on to win the Academy Award® for Best Foreign Film.

1986 For the first time in its history, the festival is scheduled for October and utilizes the Music Box Theatre. Tributes to **Sydney Pollack** and **Deborah Kerr** are presented as well as a **Claude Lelouch** retrospective and a special focus on Argentine cinema. Among the highlights are the Canadian feature *The Decline of the American Empire* by Oscar®-winning director **Denys Arcand**, while the Chilean-French cinematic master **Alejandro Jodorowsky** attends the festival to present his classic cult films *The Holy Mountain* and *El Topo*.

1987 The festival celebrates its roots, honoring Kutza's mentor **Colleen Moore** a few months prior to her death. **Peter Gardos'** *Whooping Cough* from Hungary wins the Gold Hugo for Best Film. **Brian Dennehy** wins the Best Actor trophy for his performance in **Peter Greenaway's** *The Belly of an Architect*. The festival features an amazing lineup of films by renowned directors, including **Alain Resnais** (France), **Ettore Scola** (Italy), **Juzo Itami** (Japan), **Fernando Trueba** (Spain), **Arturo Ripstein** (Mexico), **Pedro Almodóvar** (Spain), **Krzysztof Kieslowski** (Poland), **John Woo** (Hong Kong), **Paolo and Vittorio Taviani** (Italy), and **Tsui Hark** (Hong Kong).

1988 The festival honors *Midnight Express* director **Alan Parker**. Opening night features **Andrew Birken's** *Burning Secret* at the Chicago Theatre. Notable films include **Theo Angelopoulos's** *Landscape in the Mist* as well as **Bela Tarr's** *Damnation* and **André Téchiné's** *The Innocent*. Also presented is a "British Renaissance" section, which includes films directed by **Mike Newell** and **Peter Greenaway**, among others.

1989 Celebrating 25 years, the festival hosts two gala events at the Chicago Theatre: the world premiere of **Menahem Golan's** *The Threepenny Opera* (*Mack the Knife)* and a screening of the **Charlie Chaplin** classic *City Lights*. *Rocky* producer **Irwin Winkler** receives a tribute hosted by **Martin Scorsese**, and the National Film Board of Canada is honored. Soviet director **Karen Shakhnazarov** receives the Gold Hugo for *Zerograd*. **Michael Moore** attends the festival to present *Roger and Me*.

1990 Festival founder **Michael Kutza** visits Moscow for a weeklong event honoring the festival at the Russian capital's Sovinterfest. The festival has grown such that it necessitates the use of the Music Box Theatre as well as screens at the Fine Arts Theatre on Michigan Avenue. The opening night gala features the world premiere of **Lina Wertmüller's** *Saturday, Sunday and Monday* with star **Sophia Loren**, honored with the festival's Lifetime Achievement Award, in attendance. The festival also presents a 3D retrospective. Other notable films include Iranian director **Abbas Kiarostami's** *Close Up*, Egyptian director **Youssef Chahine's** *Alexandria: Again and Forever*, and Chinese director **Zhang Yimou's** *Ju Dou*, which takes home the top prize for Best Feature.

1991 The festival features a tribute to Chicago-bred actor **John Cusack**; a celebration of 20th Century Fox's CinemaScope classics; a retrospective of Spanish director **Elias Querejeta's** films; and the premiere of **Gus Van Sant's** *My Own Private Idaho*. *Delicatessen*, directed by **Jean-Pierre Jeunet** and **Marc Caro**, wins the Gold Hugo for Best Feature.

1992 Oliver Stone receives a "Director of the Decade" Award at the summertime gala. Also honored are **Jack Lemmon**, Indian director **Shyam Benegal**, Israeli director **Dan Wolman**, documentarian and director **Arthur Cohn**, and **Kathleen Turner**. The festival presents the premiere of **Alfonso Arau's** *Like Water for Chocolate*, **Baz Luhrmann's** first feature *Strictly Ball-*

room, and **Quentin Tarantino's** early feature *Reservoir Dogs*. The Education Outreach Program is launched, providing free film screenings to Chicago Public School students during the festival.

1993 The festival pays tribute to **Tom Cruise** and **James Earl Jones**. Highlights include **Robert Altman's** *Short Cuts*, Chinese filmmaker **Chen Kaige's** *Farewelll My Concubine*, and **Jane Campion's** *The Piano*.

1994 Woody Allen's *Bullets Over Broadway* opens the festival, while closing night features **David Mamet's** provocative *Oleanna*. The festival also presents a retrospective of the work of Wes Craven as well as tributes to **Rod Steiger**, Italian director **Luchino Visconti**, and **Diane Ladd**.

1995 Notable films include the premiere of **Woody Allen's** *Mighty Aphrodite*; **Bertrand Tavernier's** *Fresh Bait*; **Hou Hsiao-Hsien's** *Good Men, Good Women*; and **Marleen Gorris's** *Antonia's Line*, which goes on to win the Academy Award® for Best Foreign Language Film. Japanese auteur **Hirokazu Koreeda's** feature debut *Maborosi* takes home the Gold Hugo for Best Film. The festival celebrates the achievements of **Blake Edwards**, **Lina Wertmüller**, **Al Pacino**, and **Sally Field**.

1996 Jodie Foster receives the "Actor of the Decade" award, and audiences are treated to the premieres of **Liv Ullmann's** directorial debut *Private Confessions*, **Billy Bob Thornton's** *Sling Blade*, **Richard Spence's** *Different for Girls*, **Lars von Trier's** *Breaking the Waves*, and **Scott Hicks's** *Shine*. Also honored are **William Wyler**, **Michael Mann**, **Kim Novak**, and **Andrew Davis**.

1997 A Career Achievement Award goes to **Michael Douglas**. Also honored are **Spike Lee**, **Roger Corman**, and **Liv Ullmann**. Notable screenings include **Ang Lee's** *The Ice Storm* and **Andrew Niccol's** first feature, *Gattaca*. The prize for Best Feature goes to **Alan Rickman's** *The Winter Guest*.

1998 The careers of **John Travolta**, **Pam Grier**, **John Boorman**, and **Monte Hellman** are celebrated, while fans get a first look at **Bill Condon's**

Gods and Monsters and **Gary Ross's** *Pleasantville. Angel on My Shoulder* from director **Donna Deitch** is named Best Documentary and *The Hole* from Malaysian director **Ming-liang Tsai** is awarded Best Feature. The Education Outreach Program is expanded to provide free, year-round film screenings to Chicago Public School students and offer a special screening program for the deaf and hearing-impaired community.

1999 Patricia Rozema's *Mansfield Park* opens the festival, and audiences are treated to screenings of **Lasse Halström's** *The Cider House Rules*, **Kevin Allen's** *The Big Tease* and **Scott Hicks's** *Snow Falling on Cedars*. Tributes include **Lauren Bacall**, **Gregory Peck**, **John Frankenheimer**, **Morgan Freeman**, and visual effects innovator **Ray Harryhausen**.

2000 Lord Richard Attenborough, hometown favorite **Harold Ramis**, Asian auteur **Sabu**, and science fiction/horror director **Joe Dante** all receive tributes. **Laurence Fishburne** is honored with the Black Perspectives Award, and **Richard Gere** is presented with a Career Achievement Award on opening night when **Robert Altman's** *Dr. T & the Women* makes its debut at the Chicago Theatre. Also honored is longtime festival friend and iconic photographer **Victor Skrebneski**. The festival premieres American indie charmer **David Gordon Green's** debut film *George Washington*.

2001 Faye Dunaway's directorial debut *The Yellow Bird* opens the festival at the Chicago Theatre, and the actress is honored at a lavish celebration at the Marshall Field's Walnut Room. **Halle Berry** picks up the Black Perspectives Award at the Music Box before winning the Academy Award® later that year for *Monster's Ball*. Notable debuts include **Jean-Pierre Jeunet's** *Amelie*, **Sandi Simcha DuBowski's** *Trembling Before G-d*, and **David Lynch's** *Mulholland Drive*. **Hou Hsaio-Hsien's** *Millennium Mambo* wins the Gold Hugo for Best Feature.

2002 Clint Eastwood is honored at the Summer Gala at Navy Pier, while **Pierce Brosnan** brings glamour to opening night at the Chicago Theatre, introducing his film *Evelyn*. The year includes several seldom-screened classics, including **Harold Lloyd's** 1928 silent entry *Speedy* with orchestra and 1933's *Hallelujah I'm a Bum*, as well as **Paul Thomas Anderson's** *Punch-Drunk Love* and a tribute to actor **Charles Dutton**.

2003 Participation in the Television Awards has grown so significantly that they receive their own celebration, which is now held each April. At the festival in October, **Nicolas Cage**, **Robert Benton**, **Robert Downey Jr.**, and **Taye Diggs** are all honored. The winner of the Gold Hugo for Best Feature is *Crimson Gold*, directed by **Jafar Panahi** from Iran.

2004 Liam Neeson makes a red-carpet appearance for the opening of the festival, introducing his film *Kinsey*. Movie lovers get a first look at **Marc Forster's** *Finding Neverland* and **István Szabó's** *Being Julia* (with Annette Bening attending to discuss the film with audiences), and **Alexander Payne's** *Sideways*. **Robert Zemeckis** comes to the Cadillac Palace to introduce the world premiere of *The Polar Express* with star **Tom Hanks**. The festival celebrates the careers of **Irma P. Hall, Harry J. Lennix, Robert Townsend**, and **Robin Williams**, whose award acceptance speech consists of an hour of improvisation. In May, Cinema/Chicago presents the inaugural Chicago Youth Media Festival, screening 19 short films made by student filmmakers 21 years and younger.

2005 Opening night begins with **Susan Sarandon** greeting fans at the Chicago Theatre for the debut of **Cameron Crowe's** *Elizabethtown*. Other A-Listers in attendance include **Terrence Howard** and **Nicolas Cage**. The festival program also features **Anand Tucker's** *Shopgirl*, **Noah Baumbach's** *The Squid and the Whale*, and **Stephen Frears's** *Mrs. Henderson Presents*. **Shirley MacLaine** receives a Lifetime Achievement Award at the Summer Gala. Cinema/Chicago rebrands the Chicago Youth Media Festival as the annual Future Filmmakers Festival and screens more than 50 short films made by student filmmakers 21 years and younger in May.

2006 Tom Cruise makes a surprise appearance at the Career Achievement Award presentation for **Steven Spielberg** at the Summer Gala. The festival also honors **Dustin Hoffman**, **Liza Minnelli**, and **Ruby Dee**. **James Long-**

ley's *Iran in Fragments* is named Best Documentary, and **Asghar Farhadi's** *Fireworks Wednesday* is awarded the Best Feature Hugo.

2007 The 43rd festival is dedicated to **Roger Ebert** and opens with **Marc Forster's** *The Kite Runner*. **Tony Gilroy's** debut feature film, *Michael Clayton*, screens at the festival, as does **Anthony Hopkins's** directorial debut feature *Slipstream*. In addition, the *Chicago Tribune's* **Michael Phillips** hosts an evening honoring 100 years of filmmaking of Chicago's Essanay Film Manufacturing Company at the Music Box Theatre.

2008 Film legend **Sidney Poitier** is celebrated for his contributions to the art of film, as are **Christopher Nolan**, **Mike Leigh**, **Viggo Mortensen**, and **Jennifer Hudson**. The festival also premieres **Danny Boyle's** *Slumdog Millionaire* and **Charlie Kaufman's** *Synecdoche, New York*. Brazilian actor-cum-auteur **Matheus Nachtergaele** wins Best New Director for his film *The Dead Girl's Feast*. The Gold Hugo for Best Feature goes to **Steve McQueen's** *Hunger*, which also wins a Best Actor award for **Michael Fassbender**. Cinema/Chicago rebrands the yearly Future Filmmakers Festival as CineYouth.

2009 At the Summer Gala, the festival pays tribute to the career of **Quentin Tarantino** at the gala screening of *Inglourious Basterds*. The festival includes **Lee Daniels's** *Precious* as well as tributes to **Uma Thurman**, **Gabourey Sidibe**, **Willem Dafoe**, **Patrice Chéreau**, and **Martin Landau**. **Tina Mabry's** *Mississippi Damned* wins top honors for Best Feature.

2010 Edward Norton kicks off the opening night of the 46th Festival, introducing **John Curran's** *Stone*. The festival now regularly shows films representing more than 50 countries each year. Others saluted include **Guillermo del Toro**, **Paula Wagner**, **Ron Howard**, **Forest Whitaker**, and **Alan Cumming**. Russian director **Alexei Popogrebski's** *How I Ended This Summer* is named Best Feature and **James Rasin's** *Beautiful Darling* is chosen as Best Documentary.

2011 The Festival celebrates the careers of **Claude Lelouch**, **Martin Sheen**, and **Anthony Mackie**. Notable screenings include **Simon Curtis's** *My Week* *with Marilyn*. Finland sweeps the awards with **Zaida Bergroth** winning the top prize in the New Directors competition for *The Good Son*, and the Best International Feature prize going to **Aki Kaurismäki's** *Le Havre*.

2012 The opening of the 48th Chicago International Film Festival is perhaps the most star-studded affair ever produced by Cinema/Chicago, with **Al Pacino**, **Christopher Walken**, **Alan Arkin**, musician **Jon Bon Jovi**, and director **Fisher Stevens** all greeting fans and gracing the red carpet before a screening of *Stand Up Guys*. **Lana Wachowski**, **Andy Wachowski**, and **Tom Tykwer** visit the festival for the debut of *Cloud Atlas*. The Wachowskis reminisce about how the festival had influenced them in their formative years growing up in Chicago. **Helen Hunt** stops by to introduce *The Sessions*, and **David O. Russell** discusses his work on *Silver Linings Playbook*. **Kelsey Grammer** is honored at the festival's spring Television Awards, and the festival spotlights films from the Middle East. **Leos Carax's** *Holy Motors* takes home the award for Best International Feature, with star **Denis Lavant** honored for his performance in the film. CineYouth welcomes director **Jonathan Levine** as its opening night guest and, for the first time, presents international films as Official Selections.

2013 Opening night features **James Gray's** *The Immigrant*, and festival audiences are treated to the premieres of *Nebraska* with castmate **Bruce Dern** in attendance; *The Inevitable Defeat of Mister and Pete* with director **George Tillman, Jr.** and star **Jennifer Hudson** in attendance to discuss the film; and **Joel and Ethan Coen's** *Inside Llewyn Davis* with star Oscar Isaac. Other notable film icons who participate in discussions of their work include actor **Geoffrey Rush**, cinematographer **Haskell Wexler**, legendary Italian horror director **Dario Argento**, stage and TV actress **Elaine Stritch**, and actor **Michael Shannon**. The Best Film award goes to the Kurdish film *My Sweet Pepper Land*. The year-round Education Outreach Screenings Program welcomes more than 7,000 Chicago Public School students to free film screenings.

2014 The opening night film for the 50th Chicago International Film Festival is director **Liv Ullmann's** *Miss Julie* starring **Jessica Chastain** and **Colin**

Farrell. Honorees include **Kevin Kline**, **Gina Prince-Bythewood**, and British actress **Gugu Mbatha-Raw**. The Best Film prize goes to the Georgian movie *The President*.

2015 The opening night gala pays tribute to Italian master **Nanni Moretti's** *Mia Madre*. The festival's honorees include director **Charles Burnett**, producer **Gigi Pritzker**, and composer **Howard Shore**. The award for Best Film honors the French film *A Childhood*.

2016 The opening night premieres **Damien Chazelle's** musical *La La Land*. The honorees include Mexican filmmaker **Alfonso Arau**, director **Peter Bogdanovich**, **Geraldine Chaplin**, **Danny Glover**, Chilean filmmaker **Pablo Larrain**, **Steve McQueen**, and **James Stern**. Romania's *Sieranevada* earns the Gold Hugo for top film.

2017 The festival kicks off with a star-studded opening night featuring **Reginald Hudlin's** film *Marshall*. The honorees include **Jane Fonda**, **Taylor Hackford**, **Helen Mirren**, **Patrick Stewart**, **Vanessa Redgrave**, **Michael Shannon**, and **Alfre Woodard**. The Best Film trophy goes to the thriller *A Sort of Family* from Argentina.

2018 The opening night film is **Felix van Groeningen's** *Beautiful Boy*. The list of honorees features **William Friedkin**, **Michael Kutza**, **Carey Mulligan**, **Ruth Carter**, **Colleen Moore**, **Art Paul**, and **Paul Greengrass**. The Gold Hugo is awarded to *Lazzaro Felice* (Italy/Switzerland).

MICHAEL KUTZA – HONORS

I became a member of the French Legion of Honor in 2014,
a great personal moment. (Timothy M. Schmidt)

1962
Diplome D'Honneur
**XV Festival International
Du Film Amateur
Cannes, France**

1972
Silver Lion Award at the
Venice Film Festival

1978
Chicago Sun-Times'
**Exceptional Contribution
to Chicago** award

1985
**Chevalier de l'Ordre des
Arts et des Lettres** during
the Cannes Film Festival

1996
The city of Chicago
honorarily designated
S. Michigan Ave and
Congress Parkway as
Michael J. Kutza Way

2009
Landmarks Preservation
Council of Illinois gives its
Legendary Landmarks,
a title that is giving to
"citizens who have made
contributions to the civic
and cultural life of Chicago
and Illinois."

2010
Media Award from the
Niagara Foundation's
Peace & Dialogue Awards.

Chicago Magazine
Top 40 Chicago Pioneers

2012
American Cinematheque's
Sydney Pollack Award
honors someone who has
been of critical importance
and continuing influence in
non-profit film exhibition,
film preservation and/or
independent film distribution

2015
**Knight of the Legion of
Honour** by the President
of the French Republic for
his achievements as "an
internationally recognized
graphic designer, filmmaker
and the Founder of the
Chicago International
Film Festival."

2017
Order of Merit of the Italian
Republic, **L'Onorificenza
di Cavaliere**

2018
**Lifetime Achievement
Award** from
Cinema/Chicago

FESTIVAL JURY MEMBER

Moscow, Russia
Berlin, Germany
Jerusalem, Israel
Havana, Cuba
Tehran, Iran

Manila, Philippines
Transylvania, Romania
Guadalajara, Mexico
Taormina, Italy
Cannes, France

Catagena, Columbia
Cairo, Egypt
Los Angeles, California

ACKNOWLEDGMENTS

It took an enormous amount of time and effort to put these many years into print. I must briefly thank the individuals and institutions who were involved in bringing this project to fruition.

At the same time, I need to first acknowledge that a number of individuals who helped me so much along the way are no longer living. They include Guglielmo Biraghi, Gian Luigi Rondi, Albert Johnson, Victor Skrebneski, Linda Leemaster, Mary Ann Josh, Ray Nordstrand, Yale and Haskell Wexler, Byron Pollock, Fr. Wally, and Mary Sweeney.

For helping me begin this book project, my gratitude goes out to Robert Devendorf, John Russell Taylor, David Robinson, Lynda and Jim O'Connor, Bridget Schultz, and Jeanne Randall Malkin.

For their generous assistance and support, I must single out Dan Caliendo, Colleen Sullivan, Geno Suarez, Suzanne McCormick, John Lanzendorf, Jackie and Ed Rabin, Judy and Mickey Gaynor, Marilyn Grabowski, Vivian Teng, Mimi Plauchè, Nina Gapshis, Paula Wagner, Christie Hefner, and Abra Wilkin.

Then there are all the photographers whose work is represented in this book. They include Steve Arazmus, Robert Dowey, Matt Gilson, Stan Lazan, Dennis Minkel, Dan Hannula, Ian Sklarsky, Timothy M. Schmidt, and Tim Klein.

I also couldn't have made this project happen without the assistance of the brilliant writer and editor Ray Richmond.

I'm further sending out thanks to my legal team, including Phil Azar, Malcom Gaynor, Randy Crumpton, and Burt Kanter.

For their ongoing encouragement, I owe my gratitude to Leslie Hindman, Arthur Cohn, Donald Drapeau, John Himmel, Kim Kubiak, Candace and Chuck Jordan, Joey Mondelli, Jim Brophy, and Doug Kirk.

Lastly, my eternal thanks go out to the hardworking staff and volunteers and to all the journalists who covered the Chicago International Film Festival, as well as everyone who attended it over the years. Without you, there would have been no festival, and thus no Michael Kutza. You are all family to me.

CPSIA information can be obtained
at www.ICGtesting.com
Printed in the USA
BVHW060229011022
648155BV00003B/7

to Michael + Robert,

enjoy!

best.

Michael

10/25/22